TALES FROM BOTH
SIDES OF THE
BRAIN

Who's in Charge?
Free Will and the Science of the Brain

Human:
The Science Behind What Makes Us Unique

The Ethical Brain:
The Science of Our Moral Dilemmas

The Mind's Past

Nature's Mind:
The Biological Roots of Thinking, Emotions, Sexuality,
Language, and Intelligence

Mind Matters:
How Mind and Brain Interact to Create Our Conscious Lives

The Social Brain:
Discovering the Networks of the Mind

Conversations in the Cognitive Neurosciences

TALES FROM BOTH SIDES OF THE
BRAIN

A LIFE IN NEUROSCIENCE

MICHAEL S. GAZZANIGA

An Imprint of HarperCollins*Publishers*

HarperCollins books may be purchased for educational, business, or sales promotional use. For information please e-mail the Special Markets Department at SPsales@HarperCollins.com.

FIRST EDITION

Designed by Suet Yee Chong

Library of Congress Cataloging-in-Publication Data has been applied for.

ISBN 978-0-06-222880-2

15 16 17 18 19   OV/RRD   10 9 8 7 6 5 4 3 2 1

For the split-brain patients

Who taught the world so much

CONTENTS

FOREWORD

FROM STEVEN PINKER

S HORTLY AFTER MY ARRIVAL AT GRADUATE SCHOOL, I HAD second thoughts about whether the life of science was for me. I hadn't the slightest doubt that *science* was for me; my doubts were about the scientific *life*. As an undergraduate at McGill University I did research on auditory perception with Al Bregman, who had tied the research to deep issues in cognition and epistemology, and it was natural that I would proceed to the famed Psychophysics Laboratory at Harvard. But as I was initiated into the lab's culture, I felt the will to live draining out of me. A large fluorescent-lit room was packed with dusty audio equipment and obsolescing minicomputers, which, I was told, had to be programmed in assembly language because software packages were for weenies. The lab was inhabited by plaid-clad, pasty-faced ectomorphs, some with wives and children they rarely saw, none with a trace of humor. Their main pastime was sneering at other psychologists' lack of mathematical rigor, though they did have one indulgence: gathering around a black-and-white TV on Sunday night to watch *M*A*S*H* over pizza. The lab's first seminar, and my introduction to the dour professors who led it, was hardly more encouraging: "Let's review the latest work on Ä i over i," said one, alluding to Weber's law, the psychophysical function relating discriminable increments in a stimulus's intensity to its absolute intensity—an issue I had thought had been settled a century before, and which had inspired William James to write that "the study of psychophysics proves that it is impossible to bore a German."

Thankfully, I soldiered on, because my faith in the value of a scientific life was revived a few years later. When I was a lowly postdoctoral fellow, I was drafted to replace an ailing professor at the last minute and represent the Massachusetts Institute of Technology at a private conference in Santa Barbara, California, at which the icons of psychology George Miller and Michael Gazzaniga were to announce their plans for a new field they had christened "cognitive neuroscience." The meeting opened over red wine and antipasto on a fragrant patio with breathtaking views at the aptly named El Encanto hotel. Gazzaniga's introductory talk was periodically interrupted by wisecracks and laughter from his collaborators, and more often with wisecracks and even heartier laughter from the speaker himself. The discussion over the next day ranged from Gazzaniga's mind-boggling discoveries about the two minds housed in a split brain to speculations about how the new science would illuminate classic problems in philosophy. At the end of the day we repaired to a beautiful house Gazzaniga had built with his own hands overlooking the Pacific, accompanied by even more food, wine, and laughter, and, if memory serves, his azalea-garlanded little daughter and her friends dancing joyously in a circle. When I visualize that day I also see bluebirds and a rainbow, but I suspect they were Photoshopped into my memory by the overall impression of warmth, vivacity, and the free-ranging interests of our bighearted host.

Mike Gazzaniga is known for his monumental discoveries and for midwifing the field of cognitive neuroscience, but he is also known for showing that science is compatible with all the other good things in life. Science has its drudgery, of course, and its catfights big and small, but Mike has shown that it can be pursued with humor, friendship, sensual pleasure, and childlike curiosity. His thematic conferences, held in locales like Lisbon, Venice, and Napa and featuring two-hour presentations followed by four-hour conversations over food and wine, are a coveted alternative to the usual parade of

ten-minute PowerPoints or the warehouse of posters and their sales-
men. Nor do you have to have gray hair to be a beneficiary of Mike's
vision of enjoyable science. Mike's Summer Institutes in Cognitive
Neuroscience, known by the attendees as Brain Camp, have introduced generations of students to the field while exposing their elders
to new ideas.

The delightful memoir you are holding recounts the history of
cognitive neuroscience through the eyes of one of its founders and
most distinguished practitioners. Those who know Mike will hear
his voice in every sentence. Those who don't will learn about the
ideas, discoveries, characters, and political implications—both academic and national—of this exciting frontier of knowledge. Both
kinds of reader will be amazed by the demonstrations, ingeniously
shown in real-time videos, of the key discoveries—and isn't it just
like Mike to defy the stereotype of technology-averse oldsters and try
out a new medium of publication for the twenty-first century.

In meeting the colorful people who came and went in Mike's
life, one has to wonder how one man could be so consistently surrounded by so many people he describes as so brilliant, kind, and
funny. I will leave it to the reader to decide whether Mike attracts
such people, describes his colleagues generously, or brings out the
best in them.

Since that glorious day in Santa Barbara, Mike has certainly
brought out the best in me, teaching me, challenging me, coun-
seling me, entertaining me, and perhaps most important, showing
me that you can be a scientist and a mensch, too. And so it was a
privilege when the American Psychological Association asked me to
write the citation for his Award for Distinguished Scientific Contributions in 2008:

> *For ingenious studies of split-brain patients which illuminated the*
> *functions of the cerebral hemispheres. His discovery that the right*
> *hemisphere can act without the awareness of the left, which then*
> *confabulates a story about what the whole person did, is a classic*

of psychology, rich with implications for consciousness, free will, and the self. He created the field of cognitive neuroscience, and his accessible writings inserted it into the national conversation. His wit and joie de vivre showed generations of students and colleagues the human face of science.

PREFACE

ORE THAN FIFTY YEARS AGO, I FOUND MYSELF IN THE middle of one of the most stunning observations in all of neuroscience: the fact that disconnecting the left and right brain hemispheres produced two separate minds—all in one head. Even I, a young neophyte, understood that these unique patients were going to change the field of brain research. As it turned out, they also changed my own life to the point that I have remained a student of their secrets from that time onward. In contemplating the way to tell the story of split-brain research and how it has evolved, I have come to realize the great extent to which my own march through life has been influenced by others and how, in fact, we scientists are all a composite of both scientific and nonscientific experience. Untangling these experiences and saying which caused what is impossible. It is much better to tell the story the way it actually happened.

Most attempts at capturing the history of a scientific saga describe the seemingly ordered and logical way an idea developed. The writers of science usually do not attempt to infuse that storyline with the other realities of day-to-day living, such as the ongoing personalities of those surrounding the narrator's life. After all, scientific knowledge is what is objectively important, not scientists. While I am completely sympathetic with this view, I now realize that that approach rarely reveals what it is like to do science or to be a scientist. The raw data of measurement is one thing. Its interpretation, on the other hand, introduces the scientist and all the influences and biases working on the scientist's mind. Looking back over the evolution of my ideas, it is apparent how dramatically I've been influenced by other people. So, the actual experience in science can be quite different from the idealized view. A lot of zigzagging occurs in between the scientific experiments, as life is being lived. Science results from a profoundly social process.

The common portrayal—that science emerges from a solitary isolated genius, always laboring alone, not owing anything to anyone—is simply wrong. It is also wrong to give the budding scientist, or those who fund research, or the general public a false impression of how science happens. In this account I want to present a different picture: science carried out in friendship, where discoveries are deeply embedded in the social relations of people from all walks of life. It is a wonderful way of life, spending one's years with smart people, puzzling over the mysteries and surprises of nature. My life has been sprinkled with incredible characters, some famous; many great scientists; and some captivating split-brain patients. They all played a part in the evolution of my understanding of the overriding question: How on earth does the brain enable mind?

PART 1

DISCOVERING THE BRAIN

CHAPTER 1

DIVING INTO SCIENCE

Physics is like sex: sure, it may give some practical results, but that's not why we do it.
—RICHARD P. FEYNMAN

N 1960, MOST COLLEGES WERE NOT CO-ED. I WAS AT DARTmouth College, way out in the boonies of Hanover, New Hampshire, with hundreds of men. By the time summer came along, I had one thing on my mind. I applied for an internship at the California Institute of Technology because I wanted to spend the summer near a Wellesley girl I had met that winter. A glorious summer at Caltech, a fabled place for biology and discovery, ensued. She went on to other things. I got hooked on science. I often wonder, was I really there because of an insatiable interest in science? Or was it my interest in a girl who lived nearby? Who knows how it really works in the mercurial mind of the young? Ideas do occasionally worm their way into the interstitial areas of the hormone-addled mind.

For me, one of those thoughts was "But how does the brain make it all work?" I had also been drawn to Caltech by reading an article in *Scientific American* on the growth of nerve circuits, written by Roger Sperry.[1] The article outlined studies on how a neuron grew from point A to point B in order to make a specific connection. A lot, indeed I would say most of neurobiology, hangs on that simple question. Sperry was the king and I wanted to learn more about that. Besides, as I said, my girlfriend lived down the street, in San Marino.

It wasn't until years later when I was told about a remark made by Luis Alvarez, the great physicist from the University of California, Berkeley, that I realized that the impulse behind my question was not the same as simple curiosity. Alvarez remarked that scientists do their thing not because they are curious but because they instinctively feel something doesn't work the way they are told it does.[2] Their experimental minds kick into gear and think of another way that whatever is being discussed might work. While they can marvel at a finding or an invention, they instinctively, automatically, start thinking of alternative methods or explanations.

In my own case, I am always thinking about different ways of viewing a problem. In part, this is because of my impoverished quantitative skills. I don't find math easy and usually shy away from highly technical discussions of almost everything. I have discovered that, in many cases, it is easy to look at a seemingly complex problem using everyday language. This is true because of the way the world is. After all, one doesn't need to understand the atomic composition and quantum mechanics of the billiard ball's atoms to play a game of pool. Simple, reliable classical physics is good enough.

We humans are all constantly abstracting, that is, taking a concrete reality and from it developing a larger theory and understanding. Thus we are continually coming up with a new, simpler layer of description that is easier for a limited brain capacity to manage. For example, take my truck. "Truck" is a new layer of description for the vehicle with open space to haul stuff and which is made up of a six-cylinder engine, radiator and cooling systems, chassis, and so forth. Now that I have a new description, every time I think about or refer to my truck, I don't have to refer to all the parts and assemble them in my mind. I don't have to think of them at all (until something goes wrong with one of them). We can't deal with all the underlying complexities present in understanding the mechanisms of things every time we refer to them. It's too much for our mental processes to handle. So we chunk it—give the mechanism a name, "truck," thereby reducing its load on us from thousands or millions of items to one. Once we have an abstracted view of a previously highly detailed topic, then new ways of thinking about the topic—about how something works—become exhilaratingly clear. With the new key word and referent in hand, it is as if our minds are freed up to think again with new energy. Layers seems to be everywhere in Mother Nature.

What I will call the "layered" view of the world, which I will get back to later in the book, is an idea that comes out of the science of trying to understand complex systems such as cells, computer networks, bacteria, and brains. The concept of layering can be applied to almost any complex system, even our social world, which is to say our personal lives. One layer functions nicely, driving us with its particular reward systems. Then, suddenly, we can be bumped into another layer where different rules might apply. Caltech was going to be a new layer for me. Everything I saw and did was a "first," and there were many.

At any rate, there I was, the summer between my junior and senior years at Dartmouth, nervously walking into Caltech for my first in this long series of firsts: meeting Roger Sperry in his Kerckhoff Hall office. He turned out to be a soft-spoken, sober guy who wasn't rattled by much. I later heard that a few weeks before I met him, a monkey had gotten loose from the animal room and hopped into his office and up onto his desk. He looked up and said to his guest, "Maybe we should go next door. It might be quieter over there."

Caltech has its own heady ambience. Everyone was really smart.[3] Behind office doors were superior scientists of every type plying their trades. All universities claim that sort of thing (especially today, on their hyped-up Web pages), always extolling how truly "interdisciplinary" they are. The reality is usually quite different. But at Caltech, it was (and still is) the real deal: The engines are constantly running and then running into each other. The ethos of the place is captured by the old line, "I know he invented fire, but what has he done lately?" Working in a group that pushes you to think in unfamiliar ways is a rush. It is also challenging to keep up with the pace, to say the least. This was true for all of Caltech, and it was especially true of Roger Sperry's lab (Figure 1).

As a newcomer, I couldn't get enough of it. In retrospect, probably no one knows which parts of one's storyline account for the course one takes or explain how things turn out in one's life. Surely there are both incidental and substantive things that result in us finding ourselves in new situations and circumstances. Just as mystifyingly, in those new places, we almost instantly become part of another

FIGURE 1. The Sperry labs were on the third floor of Caltech's Alles Laboratory, close to Linus Pauling's office in the Church Chemistry Building. Across the way, in Kerchoff Hall, was A. H. Sturdevant, the father of drosophila genetics, and Ed Lewis, his Nobel Prize–winning student.

dynamic and another knowledge base. Quickly enough, we strive to achieve new goals.

It soon became evident that another interest suffusing the lab, along with nerve growth circuits—the idea that hooked me into being there—was split-brain research, which was trying to find out if each hemisphere of the brain could learn independently from the other. The place was abuzz with postdocs examining monkey and cat behavior following split-brain surgery—surgery that disconnected the two half brains from each other. Where could I jump in?

I soon came up with the idea of making a "temporary split brain." My idea was to study rats and to use a procedure dubbed "spreading depression." In this procedure, a small piece of gauze or gelfoam would be soaked in potassium and placed over one hemisphere of the brain to induce sleep or inactivity, leaving the other awake and able to learn.[4] One of the world's authorities on the phenomenon of spreading depression, Anthonie van Harreveld, had an office next to Sperry, so consults were going to be easy. He was a kind and gentle soul and very approachable, especially when it came to science. Unfortunately, that experiment never went anywhere in my hands, probably because the rats gave me the creeps!

So, I turned to rabbits. Again, the idea was easy enough. Why not inject an anesthetic into the left or right internal carotid artery, which separately supplied the blood to the left or right hemisphere respectively. That would allow me to induce sleep in one hemisphere of the brain at a time, and leave the other half brain awake and able to learn. Would it work that way? At that time in science, and especially at Caltech, the only thing standing in the way of an idea or test was one's energy and ability. No IRB (Institutional Review Board), no lack of funds, no discouraging cant from others, no endless regulations. You could just do it.

I had to have a measure of neural activity to make sure the appropriate half brain was asleep while the other was awake, so I began by pulling together an electroencephalograph, that is, an EEG. Then I had to learn how to teach a rabbit a trick so that it would learn something. We decided to teach the rabbit to blink an eyelid to a sound. I got that under control. Then I had to learn how to attach recording electrodes on the small rabbit skull in order to pick up the electrical activity, the EEG. I somehow managed to do that. Finally, I had to be able to inject into either the left or right internal carotid artery (the key arteries leading from the heart to the brain) an anesthetic and be convinced the drug stayed lateralized to half of the brain and didn't leak over to the other half brain and thereby put it to sleep as well. After a lengthy library search on the anatomy of the Circle of Willis, the arterial structure at the base of the brain, I decided it would work in the rabbit. Even though it seemed that blood from both sides of the supplying arteries would mix at the Circle, some studies showed that, due to a special hemodynamics, it didn't. I went ahead, convinced that hemodynamics would save the day and hoped an anesthetic applied to one carotid would stay long enough in one half brain for me to do the experiment. At last, I was ready to rock and roll.

My laboratory space for all of this was in the hallway in Sperry's lab area. Space was tight, as there were many active postdocs toiling away with their own studies. One day I was busy doing a trial run. All the components were in place: the rabbit, the EEG machine recording the neural activity and spitting out the results on paper, and the eight ink pins fluttering back and forth. Then Linus Pauling walks by. Now, everybody knew who Linus Pauling was, particularly in our building, as his office was just around the corner in the chemistry building. He was one of the founders of quantum chemistry and molecular biology and was ranked as one the most important scientists of the twentieth century, making his way in 2000 onto an American stamp. Pauling stops and asks me what I'm doing. After sizing up the situation he says, "You know those squiggles you are 'recording' may be nothing other than the simple mechanical consequences of a Jell-O-like substance in a bowl. You ought to test that first."[5]

As he walked on down the hall, I caught the fever. His message was simple: Assume nothing, young man, and test everything. No matter where you turned, people were challenging, questioning, poking, yet encouraging, and yes, supporting the notion that it might work differently, which urges the young scientist on. It was intoxicating. Little did I know that a couple of years later, after winning his second Nobel Prize, Pauling would be suing for libel William F. Buckley Jr., who was about to become my lifelong friend!

It was in this setting that a little over a year later, I tested the first split-brain patients. I wanted to see what people were like who for medical reasons had had the two hemispheres of their brain were surgically separated: the left brain no longer connected to the right brain. This book is about what that particular medical reality is, means, and taught us. The biographical details about the many scientists both directly and indirectly involved, who are featured in this telling, have been pruned from previous, mostly purely scientific accounts. As I have reflected on my own body of research, I feel it is important to grasp at least one tale of how many seemingly unrelated experiences flow together to make a life, and in this case, my life in science. But I am getting ahead of myself.

Over that all too short summer, I got the rabbit preparation to work. There was constant kibitzing from others in the lab, but the task I chose was mine to do. The notion of discovering a little bit of the way something worked was palpably exciting. I was seduced. I knew then I had to discuss this with my father. His dream was for me to follow in his and my brother's footsteps and attend medical school. My father was a force. Breaking away from the padrone's plan required a conversation.

ORIGINS

Dante Achilles Gazzaniga (Figure 2) was born in Marlboro, Massachusetts, in 1905. After attending St. Anselm's College in Manchester, New Hampshire, he was bound for home to work at the boot factory where his father had worked since emigrating from Italy. The local priest intervened, the same one who had been instrumental in getting him to college. He told my father that if he studied chemistry and physics over the summer, he would arrange for him to go to medical school at Loyola, in faraway Chicago. Oh life was so simple and straightforward in those days. Learn the stuff and you get to go to the next step. And that is what he did. He went to Chicago in 1928, and with the money his mother had saved, he planned to buy himself a microscope. Unfortunately, the money was in the bank, and it was all lost in the 1929 crash.

FIGURE 2. Dante Achilles Gazzaniga dropped everything he was doing in Los Angeles to join the U.S. Navy and serve during World War II. He offered soldiers surgical care at bases in the New Hebrides and New Caledonia.

In Chicago, he lived around the corner from where the infamous Valentine's Day Massacre occurred, carried out by the gangster Al Capone. He even heard the shots on Clark Street. My father would sometimes get clam chowder at a local dive right by the alleyway where the shooting took place and sneak out packages of oyster crackers, which were a main part of his diet. To support himself and pay tuition, he played semipro football, as he was tall and strong, and he also ran an elevator, in which he did much of his homework. Somehow he got it done and I have thought about how different our experiences were, as I had enjoyed a paid research assistantship in luxurious Pasadena, California.

After four years of Chicago, he headed off to the train station with a plan: get on the first train heading to a sunny place. Successful in this goal, he stepped off in Los Angeles, where he did an internship at the famous County Hospital in 1932–33. Headed to the Rose Bowl game with his buddies, he was trotting down the hospital's front steps on New Year's Day 1933 when he first met my mother, who was on her way in to work. Three and a half months later, they were married. At one point in my mother's lively life, she was the secretary to the famed Aimee Semple McPherson, the evangelist who founded the Foursquare Church and captured the imagination of Los Angeles with her sermonizing at the Angelus Temple she'd built. It may well have been that my mother's famous father, Dr. Robert B. Griffith, had landed her the job in the media-conscious town. He was the first plastic surgeon in Los Angeles and a hugely talented and successful physician. Among his patients were Hollywood stars including Mary Pickford, Charlie Chaplin, cowboy star Tom Mix, and Marion Davies.

My mother's father, whom I never met, was also known in local circles as a great chess player (masters level) and was a very good friend of Herman Steiner, the longtime chess columnist for the *Los Angeles Times*. They were both on their way back to Hollywood from a chess match in 1937 when they were hit head-on by a drunk driver. My mother found out that her father died in a car crash by reading the newspaper. I recently saw a picture of my grandfather for the first time and noticed some similarity in our facial features, though the chess gene didn't find its way to me. (My brother Al got that.)

Life in Los Angeles was fast-paced and colorful, but it was the Depression, and jobs were tight, even for physicians. Unable to find a job in Los Angeles, my father found work as a doctor for the men building the Colorado River aqueduct, which channeled water all the way through Arizona to California. It was a vast project. Nonetheless, in his spare time in the desert, my father had other projects going. He prospected and set up mining claims all over the place but gave them all up to the government years later when he signed up to join the World War II effort. My father always had multiple activities going at the same time and he worked at all of them. All of his children adopted the trait.

My father's cousin, who was a doctor in North Adams, Massachusetts, drowned. The family called my father to move back, so he, my mother, and their new baby, my oldest brother, Donald, climbed into the family car, a DeSoto sedan, and made the trip to North Adams in the summer of 1934. They were put up in a house a ways out of town. During snowstorms when my father was stuck in town, my mom, the original California girl, was isolated in the sticks hovering in front of an open oven with the baby to keep warm. Meanwhile, my father was hanging out and playing cards with the boys in town. That didn't wear well. The following February, in the depths of a western Massachusetts winter, my mother's cousin sent her a sprig of orange blossoms from sunny California. That put her over the edge. My father didn't like the weather, either, so they moved back to Los Angeles after about nine months. He hooked up with the Ross-Loos Medical Group, which was just starting up, and became one of the founding partners. That medical group became the first HMO in American history and served as the model for the now-huge Kaiser Permanente.

Clearly, my father was full of spunk and a bit of a maverick. His roundabout course to professional success was objectively evident to me, but did he see it that way? I didn't know what to expect when I broached the news about my new plan. "Dad, I think I want to go to Caltech rather than medical school." There. It was laid out clean and clear. My dad looked at me with deep medical authority and said, "Mike, why would you want to be a Ph.D. when you can hire one?" He was truly puzzled. My father was committed to medicine like few others and was there to serve the sick. I can remember more vacations either canceled or shortened than enjoyed because patients always came first.

Nonetheless, after a moment, Dad smiled and wished me luck. After all, there was still the small matter of being accepted into Caltech. The description of what a student had to be like in order to even be considered by Caltech bore no relation to me. As I have already mentioned, the place was chock full of mighty smart cookies and most of them could run circles around me. I had come to learn, however, that a large number of students were there for another reason: They had somehow proven to their future mentors that they knew how to do stuff. This usually came about by serving in summer fellowships just like I had done. This would be my only hope of getting in.

THE UNDERGRADUATE LIFE

Sperry came to bat for me. He had been impressed with my rabbit work and overall energy, and the following spring, my senior year at Dartmouth, the Caltech biology department conditionally accepted me to graduate school. Clearly, I had to show my stuff the first year.

It had been a challenging four years at Dartmouth. Little did I know, however, that due to my membership in the infamous Animal House (Figure 3), my social life there would become a more noteworthy achievement than anything I had ever accomplished academ-

FIGURE 3. The Alpha Delta Phi (Animal) house at Dartmouth College. A few years back, some of us former "animals" met for a reunion. It didn't take long for us to decide that the place should be bulldozed.

ically. A science geek among the more notorious animals, I played out my days as "Giraffe." I was the nerd of the fraternity, preferring to spend more of my time working in the laboratory of the psychologist William B. Smith than drinking in the Alpha Delta Phi House basement.

Smith had a passion for research. He had built a small lab on the top floor of McNutt Hall, where we developed methods to measure eye movements. We worked together long into the night. Research was all new and exciting for me, and the first tantalizing glimpses of the search for one of Mother Nature's mysteries had me hooked. Yet, at the time, before that momentous summer at Caltech, it simply seemed as if it were another thing to do in order to get into medical school. I did make some of my dearest friends at Animal House, and the tone of the place motivated me to get on with life!

So, during my senior year, while my Dartmouth days were winding down and my serious itch for Caltech had taken hold, I became captivated by the question "What would happen to humans with a corpus callosum section?" (*Section* here refers to surgery cleaving the brain's biggest nerve bundle.) It was clear after my Caltech summer with the rabbits' brains and the heavy emphasis on basic research that I would move toward both. At the time it was inconceivable to think that humans might show the dramatic disconnection effects that were being shown in animals. No one really thought that a human being with an object placed in his or her left hand would be unable to find an object to match with the right hand. That just seemed nutty.

In the tradition of Francis Bacon, it was time to count the horse's teeth. This possibly apocryphal story captures what science is all about:

> In the year of our Lord 1432, there arose a grievous quarrel among the brethren over the number of teeth in the mouth of a horse. For thirteen days the disputation raged without ceasing. All the ancient books and chronicles were fetched out, and wonderful

and ponderous erudition such as was never before heard of in this region was made manifest. At the beginning of the fourteenth day, a youthful friar of goodly bearing asked his learned superiors for permission to add a word, and straightway, to the wonderment of the disputants, whose deep wisdom he sore vexed, he beseeched them to unbend in a manner coarse and unheard-of and to look in the open mouth of a horse and find answer to their questionings. At this, their dignity being grievously hurt, they waxed exceeding wroth; and, joining in a mighty uproar, they flew upon him and smote him, hip and thigh, and cast him out forthwith. For, said they, surely Satan hath tempted this bold neophyte to declare unholy and unheard-of ways of finding truth, contrary to all the teachings of the fathers. After many days more of grievous strife, the dove of peace sat on the assembly, and they as one man declaring the problem to be an everlasting mystery because of a grievous dearth of historical and theological evidence thereof, so ordered the same writ down.[6]

The teeth, in my case, were the human patients at the University of Rochester who had undergone surgery similar to what had been done on the animals at Caltech. In the early 1940s, this famous group of patients had had the corpus callosum sectioned in order to limit epileptic seizure activity to one half of the brain. This process split and disconnected the two hemispheres of the brain.

The surgery was performed by the neurosurgeon William P. Van Wagenen, who had noticed that an epileptic patient who had developed a tumor in the corpus callosum was having fewer seizures. He wondered if severing the corpus callosum would stop the spread across the brain of the electrical impulses that elicited the seizures. So he sectioned the corpus callosum in a series of twenty-six patients with severe uncontrollable epilepsy. Seemingly well examined by a young and talented neurologist, Andrew J. Akelaitis, these patients had a remarkable decrease in the number of seizures they experienced, without any major behavioral or cognitive changes following surgery. Disconnect the two hemispheres and nothing appeared to change! Everyone was happy. That finding sat in the literature for ten years. Karl Lashley, the leading experimental psychologist of the era and Sperry's postgraduate advisor, had seized on this finding to push his idea of mass action and the "equipotentiality" of the cerebral cortex; he claimed that discrete circuits of the brain were not important, only its cortical mass.[7] Quoting Akelaitis's work, he concluded that cutting the massive nerve bundle that connected the two halves of the brain appeared to have no effect on interhemispheric transfer of information and quipped that the function of the callosum was to keep the hemispheres from sagging.[8]

The Akelaitis patients, as they were called, seemed like the perfect patients to confirm or refute whether the Caltech animal work done by Sperry and his graduate student Ron Myers applied to the human brain. It was then known from the animal work that after splitting the cerebral hemispheres, the monkey's left hand didn't know what the right hand was doing. Could that possibly be true for humans? While it seemed crazy, I was convinced it had to be. I wanted to retest the Rochester patients.

I figured out who might know about those patients in Rochester and made a call. It worked and through the offices of Dr. Frank Smith, who at the time of the surgeries in the early 1940s had been a resident and operated on the very patients in question, I was going to be allowed to see the patients, if I could find them.

I designed many experiments that were different from those done by Akelaitis and exchanged letters with Sperry about the ideas and the plan. I applied to the Mary Hitchcock Foundation at Dartmouth Medical School and received a small grant (two hundred dollars) to rent a car and to pay for my stay in Rochester. I drove out to Rochester and went straight over to Smith's office to begin sorting through his files to find possible names and phone numbers. While I was there, he called to say that he had had a change of heart and he basically asked me to bug off. Even though my car was loaded with borrowed tachistoscopes, pre-computer-age devices that display images on a screen for a specific amount of time, and other paraphernalia from the Dartmouth psychology department, I departed as requested. The effort to reveal the effects of a human corpus callosum section was left for a later time.

A few months later, however, I was back on the road, and rather than being disappointed, I was excited. I was headed to Pasadena. For five glorious years, Caltech was to be my home.

DISCOVERING CALTECH

It was a grand adventure to go from the Animal House to the so-called J. Alfred Prufrock house, across the street from the Caltech biology building (Figure 4). Helping me to get settled was one of Sperry's senior graduate students at the time, Charles Hamilton, who soon became my best friend there, and had urged me to live at the Prufrock house. By the time I got there, it had a huge reputation for smarts, for parties, for just about everything. Chuck's roommates, who already graced the rented two-story home, included Howard Temin, who went on to win the Nobel Prize for his groundbreaking work on viruses, and Matt Meselson, who coauthored with Franklin Stahl one of the most famous experiments in all of molecular biology.* When I moved in, Sidney Coleman and Norman Dombey, two theoretical physicists—one studying with Richard Feynman, the Nobelist and celebrated populizer of science, and one studying with Murray Gell-Mann, another Nobelist, who coined the term *quark*—were living there. Coleman went on to a distinguished career at Harvard and became known as the "physicist's physicist."

The weekend parties at the Prufrock house were of a different cali-

---

* Their work supported the hypothesis that DNA replication was semi-conservative, using one strand of the original DNA helix and one newly minted one during replication. M. Meselson and F. W. Stahl, "The Replication of DNA in Escherichia coli," *PNAS* 44 (1958): 671–82.

FIGURE 4. The so-called J. Alfred Prufrock house was a fabled place for graduate students to live at Caltech. My roommates Sidney Coleman, Norman Dombey, Charles Hamilton, and I threw lots of parties.

ber than those at the Animal House. At one such party, Richard Feynman showed up. As he was leaving, Feynman came up to me and said, "You can split my brain if you can guarantee I can do physics afterwards." Laughing, I said, "I guarantee it." Quick as a flash, Feynman stuck out both his left and his right hand to shake on the deal!

Margaret Mead once remarked that she thought all Caltech men thought women had a staple in their belly button because the only time they viewed a women naked was in the foldout of *Playboy*. She was rough on them, and the student newspaper in April 1961 called her out:

> *Tuesday night, to an overflowing audience, Dr. Mead explored the question of "The College Man's Dilemma: Four Years of Sexual Uncertainty." With several barbs for the Techmen in the crowd, she considered the kind of culture in which Caltech exists and some possibilities for improvement. This culture, said Dr. Mead, believed that sex is really necessary for health. This attitude has led to early marriages, which are, according to Dr. Mead, incompatible with the development of the highest mental facul-*

ties. Her talk implied that perhaps Techmen should not marry until much later, if at all.[9]

The mystique of the Caltech undergraduate life remains today, featured in the TV series *The Big Bang Theory.*

As a graduate student, I got to know many of the undergraduates and many remain fond friends today. For example, I came to know Steven Hillyard at Caltech, as he took an early interest in split-brain patients and is by far one of the best scientists I know. He lets the data do the talking and is a stickler for details. Steve and I have collaborated over many years, and to this day we remain in constant touch. His quiet demeanor masks a penetrating intellect and a firm sense about what is going on in any chaotic situation, whether it be a pile of scientific data or a barroom full of drunks. This skill has enabled him to produce a string of talented students, all highly successful. He set the benchmark.

Harvard, Stanford, Caltech, you name it, all have prestigious graduate schools in science. One unheralded fact in academic life, however, is that most graduate students couldn't get into the undergraduate part of their graduate school. While there are always exceptions, such as my housemates in the Prufrock house, this tendency suggests that most privileged undergraduate schools don't send their students off into science. Law schools, medical schools, and business schools seem to grab most of the students from the top schools. At Caltech, graduate students are smart, but astounding differences frequently occur between graduate students and the fabled undergraduates.

As soon as I arrived for my first day of graduate work, Sperry gave me my assignment. I was to implement the split-brain experiments I had designed with him during my senior year at Dartmouth, but on Caltech patients rather than Rochester patients. Before I knew it, I was in the thick of an exciting and consuming project, examining a robust and charming man, W.J., who was about to undergo cerebral commissurotomy, the so-called split-brain surgery, to control his otherwise capricious epilepsy. He was the sort of levelheaded person to instill respect, especially in a young, green graduate student like I was.

Joseph Bogen, a neurosurgical resident at the time, had critically reviewed the medical literature and was convinced that split-brain surgery would have beneficial effects. It was he who had launched the project. He enlisted Dr. Philip J. Vogel, a professor of neurosurgery at the Los Angeles–based Loma Linda Medical School, to perform the operation. My chore was to quantify the psychological and neurological changes, if any, in the way W.J. behaved once the connections between his hemispheres had been severed.

The conventional wisdom suggested that nothing would happen. As I have already mentioned, twenty years earlier Andrew Akelaitis had found that cutting through the corpus callosum in human subjects produced no behavioral or cognitive effects. It fell to me to test W.J. I was the luckiest man on earth.

As best as I can figure out, luck is a big part of a life in science. Most people have the intellectual horsepower to do science, and most scientists are smart people. It is also true that most academic scientists toil at their fields, making contributions, teaching their courses, and living fulfilled lives. Some, however, get lucky. Their experiments reveal something not only interesting but important. The spotlight falls on them for a while, and they either revel and enjoy it or simply accept it and continue on their way in hopes of doing something else of interest.

Sperry had more luck than most. For example, in the early 1960s, the histology technician Octavia Chin apologized to Roger because she couldn't get the regenerating fibers of gold fish to stain the same color as normal fibers. Just then, Domenica "Nica" Attardi, a young Italian postdoctoral fellow, came in asking for some part-time work. Nica took on the question of why the fibers didn't stain and there followed an elegant study by Attardi and Sperry[10] of the pathway taken by a regenerating axon in the fish visual system, which became a classic example of Sperry's ideas on neural specificity. Pure serendipity. I know this kind of thing happens, as I have come to experience it in my own life on several occasions.

Once I began my graduate work the days were long and electrifying. One time I got home late, around four in the morning, and I noticed Sidney Coleman's light was on. There he was lying on his bed, staring up at the ceiling. I asked him what was up. Sidney barked back, "Shut up! I am working." Newly appreciative of the gap between physicists and biologists, I once asked Norman Dombey what he was thinking about when he walked around the house with a somewhat dazed look on his face. "Oh," he said, "I am usually wondering if there is a Coke in the house."

Even back in those relatively simple days, the normal nine-to-five workday became hectic, way too short, and endlessly interrupted, and so the work stretched late into the night. To solve the problem, I took to going to work at midnight and going home the next afternoon to sleep at six. The nights were wonderful times to work, no interruptions, time to think and time to build the new devices I needed. I kept this schedule for a long time.

Another one of the many things I had learned was the importance of staff. Everyone used to joke how the dishwashers for the molecular biology labs would come in on the holidays and weekends if a graduate student needed them. It was true. Everyone had some version of the fever. After all, Meselson and Stahl had just carried out their famous experiment, and Howard Temin was being launched by Caltech's Renato Dulbecco* and beginning to work on viruses. Throw Bob Sinsheimer, Max Delbrück, Ed Lewis, Ray Owen, Seymour Benzer, and a dozen or so other world-renowned molecular biologists into the mix and you can begin to get a sense of the place.

---

* Delbucco, from a small town in Calabria, Italy, was a virologist who won the 1975 Nobel Prize for his work on oncoviruses, which are viruses that can cause cancer when they infect animal cells. He had been a member of the Italian Resistance during World War II before moving to the United States.

I had discovered the importance of the shop technician, Reggie, when he helped me make my animal training device. The backbone of the Sperry lab was another technician, Lois MacBird, who prepared everything for surgeries, among other chores, such as running the whole show. The senior postdoctoral fellow at the time, Mitch Glickstein, recently reminisced, "Lois was the steady bedrock of technical help. She trained monkeys and prepared and assisted in surgery. Sperry never reproached people, he needled them. Harbans Arora, a research fellow who had trained at a fishery in India, had very little ability to tell when Sperry was teasing. Sperry came in while Harbans was operating and noted that his white surgical gown was not color coordinated with a green surgical pack. Not realizing that Sperry was teasing him, Harbans found Lois after the surgery and said, 'Lois! You must never put a white gown in to sterilize with a green surgical pack. Roger was very angry.'"[11] Lois had a wonderful ability to smile this off and life went on.

Of course, it was people like Mitch who really made the atmosphere intoxicatingly different. The postdoctoral layer is crucial in scientific training. Postdocs arrive at a lab, already deeply knowledgeable about some aspect from the science at hand. Swooping up the graduate neophytes, the postdocs offered not only intellectual but social aid. Mitch, a student from Boston Latin High School and University of Chicago, was eager to share his deep sensitivity about life, both the work and the fun. We used to steal off together during the week and go to the horse races at Hollywood Park and Santa Anita. Among the many things Mitch taught me was the racing form.

Joe Bogen was also in this category. Yet it was difficult to think of him as a postdoc, as Joe was a neurosurgical resident, a real medical doctor, who had spent time at Caltech as a postdoc but was now fully immersed in his medical-surgical training at White Memorial Hospital, then affiliated with Loma Linda University. Joe and his terrific wife, Glenda, brought a rare, exuberant gusto to the more sedate Caltech. I was always going over to their apartment for dinner and discovered the trick of having a bottle of frozen vodka in the freezer. Left-wing politics were always being discussed, which I enjoyed, even though my leanings were growing conservative at the time. He used to talk about his father, a lawyer, who Joe said was famous for the Bogen line at the draft board. He said that his father had won a landmark case on a conscientious objector who claimed he had never taken the oath to serve. After Bogen's father proved his point, the Selective Service made recruits physically take a step forward across the "Bogen line" to prove their commitment. It's one of those stories that are too good to check.

With all of this richness and activity, the unquestionable driving force behind the lab was Roger Sperry (Figure 5), or Dr. Sperry, as we all called him. He was both elusive and omnipresent. He could be aloof, such as when he wouldn't come out of his office to meet Aldous Huxley, or utterly engaged with a lesser mortal who seemed at sea to others. Soft-spoken, yet prodding the status quo in so many ways, he was not reluctant to needle his rivals. After one of his lectures, a particularly aggressive questioner wound up with a whimsical stare from Sperry, who then simply said, "Boy, it sounds like you got something going for you."[12] And then he turned away.

Upon my arrival for my graduate years, I started studying patients and immediately began to spend approximately two hours

FIGURE 5. Roger W. Sperry was the inspired leader of the Caltech program in psychobiology. He was one of the forerunners in neurobiology research, which changed the way many scientists thought about brain development. He went on to develop the psychobiology program at Caltech.

a day speaking with Sperry, a habit that lasted throughout my stint at Caltech. We talked about everything. After my frequent solo trips to the patient's home for testing, I always came back to give a full report in debriefs that could last as long as the actual testing session. Sperry always took copious notes, and it was obviously a time when our ideas became mixed and fortified. I was the novice and he was the pro. But because he was not yet a pro in this new field of human research, I also served as his scout. Together we hashed things out in dozens upon dozens of such meetings. Glickstein claims that I am the only person alive who could ever get Roger to smile. While I am not so sure about that, we did have a wonderful relationship that was largely built during these sessions. James Bonner, the distinguished biologist, once quipped, "Maybe we should keep Mike around so Roger can have someone to talk to." It was easy for me, as I was devoted to the work, the man, and his mind.

Of course the memorable peaks of life come scattered among the many hard and often dreary days of work. On one bright Sunday afternoon, Steve Allen, whom I'd gotten to know, brought his entire family over to the lab to see exactly what we do. Steve, who became a lifelong friend, was like that: utterly unassuming, endlessly curious, and always positive—like Tom Hanks, he was considered one of the good guys in Hollywood. His family was suitably intrigued and polite. At the end of the visit Steve asked, "What percent of the work is exciting?" After thinking for a moment, I replied, "Oh about ten percent. The rest is routine." As I have learned in life, 10 percent is a good number for most professions. I know it has been enough to keep me going to work every day with a smile on my face.

It was the occasional meeting of public figures like Allen that slowly made me realize that nonscientists want to know about basic research, too. Back in the 1960s, "outreach programs" were nonexistent. The Ivory Tower mentality dominated intellectual discourse, and as a result, the natural social isolation of researchers only intensified the two cultures. When Steve, one of the top comics of the day, wanted to know more about the fibers of the corpus callosum, it started to become clear to me that public communication of science is a good thing, so long as it is done with accuracy.

In recounting the past, we tend to concentrate on the positive times. There were plenty of negative experiences, but I didn't dwell on them. Aside from the hugely disappointing emotions that accompany the failed experiment, the useless finding, or the bumbled test, there is always personal conflict, such as academic bullying, in science. For the life of me, I don't know why, but smart people like to point out how stupid someone else seems to be. The common belief is that greater education leads to a greater tolerance and appreciation of human individual variation. If only it were true. People are constantly flexing and showing their prowess and absolutely love one-upping each other. Take Max Delbrück.

Delbrück was a legendary figure at Caltech and remains, deservedly, an icon in the history of biology. While his own research was of high quality, his fame was really based on his critical powers. It's commonly said that during the heyday of molecular biology, not a single noteworthy paper was published unless Delbrück approved of it.

The event where people showed off was the weekly Caltech biology seminar. Max would always sit where he could be seen and not let anything slip by. Among his many skills, Mitch Glickstein is a superb historian of neuroscience and recounts a typical scene when he was challenged.

When I first got to Caltech I was urged to give a seminar. As a psychology student I knew very little of interest, but I had worked in the Kleitman lab for a year, and I spoke about REM sleep. I made a fourfold table: REM/not REM Dream reported/not reported. Max got up immediately and said "Oh no that's wrong." I looked again and said "That's right," whereupon he said, "Oh yes, that's right."[13]

In my experience, the tough guys are not tough all the time. Max, for example, took students and fellows out camping at Joshua Tree National Park. Max loosened up on these trips, and they tended to be full of wit, knowledge, and adventure. Invitations were coveted, and everyone always came back raving about the experience. The social psychologist Leon Festinger once told me that in order to keep the French Foreign Legion in line, they only had to shoot a few deserters, not three hundred. A periodic bit of nastiness might go a long way to keeping the ship on a straighter line and everybody on their toes.

POLITICAL ADVENTURES

For me, a life in science was never all science. While consuming, it wasn't all-consuming. There are other personal needs, like income, politics, relief from the anxiety that one might only be meeting low expectations in the lab. As a consequence, the lowly beginner's role found me involved in all kinds of other activities. One day someone suggested I could earn some extra money as a graduate student by taking on directorship of the office of graduate student affairs at Caltech's spanking-new Winnett Student Center. The job came with an office, a secretary, and a small salary. I snapped up the opportunity, which I thought would nicely play into various projects I was launching. Oddly, I can't remember a single thing I did under the aegis of this office. I had a very pleasant secretary who ran the day-to-day functions, but what were those functions? No clue. They must not have meant much to me. At the same time, I was learning how necessary it was to piece together ruminative outside projects in order to pay all the bills when working on an academic salary.

I did, however, engage in other outside activities, and they were truly bizarre, given my chosen line of work. In my senior year at Dartmouth, I exchanged letters with a Jesuit priest who worried about me and my stirrings and misgivings about parts of Catholicism. He kept hammering on the point not to be mad at the church, because all of us are the church. The arguments didn't work and over time I lost my faith.

What I found straining in graduate school was the uniform commitment to secular liberalism and to its insistence that social justice should mainly be achieved by the state. I'd inherited from my father a Catholic sense of social justice, with its belief in the dignity of work, family, responsibility, and aiding the poor. Catholic social justice and secular social justice have a lot in common, even though the views are derived from different core beliefs. In short, my nascent questioning of my own social and political assumptions was on the move. My exuberant college views that everything could be fixed and, if not fixed, forgiven were falling apart. The secular insistence that social services could fix all that was broken led me to believe that liberalism was a cruel hoax. At the time, brains appeared not to be as mutable as the liberal activist wanted them to be. I was also beginning to doubt fancy psychological theories of development and to become convinced that it was almost impossible to change anybody's behavior in a serious way. My thoughts were surely a mishmash of my newly found knowledge about brains, how they are wired up in specific ways, and the desire most of us share to fix people and institutions that were dealt a bad deck of cards. These primitive urges provoked me to try to learn more about politics and other ways one could spend one's life.

So, some friends and I started something called the Graduate Committee for Political Education. We were tired of all the liberal speakers who were routinely invited to Caltech. Where were the conservatives? We knew Caltech wouldn't come along quickly or quietly, so we started our own outside group, rented a public auditorium in nearby Monrovia, and arranged for the young enfant terrible on the right, William F. Buckley Jr., to give an evening lecture. Buckley was the brash editor of the new conservative magazine *National Review* and someone who could stir the pot with wit and a touch of irreverence. My two other friends, sassy lawyers from Harvard who had jobs in Los Angeles, and I thought we were pretty cool, even

FIGURE 6: Senator Barry Goldwater visited Pasadena and hinted that he would promote our first lecture by a conservative.

bizarre. But, once committed, we worked it hard. When Barry Goldwater visited Caltech I was introduced to him and asked him if he would agree to help promote the lecture event. He did (Figure 6).

I met Bill the day before the talk at the home of his sister-in-law, who headed the local Red Cross chapter and lived in Pasadena. It was a poolside lunch with, I will never forget, onion sandwiches. Now, have you ever had an onion sandwich? Bill was quick to put me at ease, even at his boyish age of thirty-six. We chatted about anything from his sister-in-law's sandwiches to John F. Kennedy. I remember using the word *potentiate,* which is a common one in pharmacology, and him informing me that no such word existed in the English language. That was the last and only time that I was right in a dispute between the two of us that had to do with language.

That weekend, a friendship was born that survived more than fifty years. Once again I learned that nonscientists wanted to learn more about science. While I wanted to know about politics, he wanted to know about brains, about drug use, about computers, about what was being discovered about life! Little did I know then that over his lifetime I would be one of his contacts, his scouts for scientific knowledge. While I was enthralled with every tidbit he uttered about politics, he wanted conduits to scientific thinking and I gave him these.

Bill was naturally friendly and unflaggingly generous, though I believe he had no concept of the many implicit gifts he gave his friends. Most of my close friends are in science, which is to say that they reflexively try to dissect assumptions on scientific claims. Yet, as a group, they are not prone to applying those skills to social and political agendas, let alone doing it with wit. Bill challenged everything, but always with a grin and with humor. His was a disposition that made it hard for others to rattle his resolve. He always was on top of things with the big picture. Expressing that attitude about life served those who knew him in ways that I don't think he ever fully appreciated. It surely influenced how I dealt with my academic friends the rest of my life. I learned that holding a minority view can be fun, and that if it is done in good spirit, those around you can have fun as well. Overall, Bill was a risk taker, yet prudent and mannered. He once told me he didn't like to meet people he admired because they invariably disappointed in person. Gregarious yet private, Bill never disappointed.

Soon after the lecture in Monrovia, I discovered that there was a bit of Sol Hurok* in me. A couple of weeks after that evening's great success, we decided to go big time. Why not arrange a series of debates on the American Constitution? Why not put out a book?[14] Why not have some fun? So I asked Bill if he would lead off such a series debating Steve Allen on the American presidency. He said, "Sure." Then I asked if he would write to Steve Allen, since I didn't know him yet. "Sure," he said, adding that Allen's wife, Jayne Meadows, had grown up in his hometown. Bill wrote the letter, Steve said yes, and within a couple of weeks, I had arranged for two other debates. I had Robert Hutchins, former president of the University of Chicago, a post he attained at the age of thirty, debating Bill's brother-in-law, L. Brent Bozell, another lawyer and ghostwriter for

---

* Sol Hurok was a world-famous twentieth-century American impresario who managed Arthur Rubinstein and Isaac Stern, among a fleet of other well-known actors and musicians.

Barry Goldwater's *Conscience of a Conservative*, about the Supreme Court. Finally, somehow I had arranged for James MacGregor Burns, one of JFK's biographers, to debate Willmoore Kendall, the maverick conservative political theorist who had been fired by Yale. Their topic was the Congress. I don't know what I was thinking. A few weeks later, I realized I had signed contracts for auditoriums and speakers that totaled more than ten thousand dollars. The Graduate Committee for Political Education had two hundred dollars to its name.

By the morning of the first debate, which was to be held at the huge Hollywood Palladium, only two hundred people had purchased tickets, some of which had been peddled by my little sister at her junior high school. Steve had taped his TV show the night before with Bill as a guest. They had warmed up for their debate about JFK, but the show wouldn't air for two weeks and wouldn't aid in ticket sales. I was concerned and told Steve. Steve very matter-of-factly said, "Don't worry, Mike—three thousand people would show up to watch me play tiddlywinks." I wasn't convinced. On the way to the event, we had stopped by the house of my wife's friend, who was in the restaurant business. I had met my wife, Linda, through Sperry's student Colwyn Trevarthen, and his wife, who came from a longtime Pasadena family. Linda also had been raised in Pasadena. Her family knew the business community, and she was close to many of them. Linda's friend asked one question: "How are you fixed for change?" Not only was I not fixed for change, but it soon became clear I had no idea what I was doing. He intervened, grabbed his wife, went down to their restaurant, gathered up several hundred dollars of quarters and dollar bills, and helped man the ticket booths at the Palladium. As it turned out, three thousand people bought tickets that night, and two of them were Mr. and Mrs. Groucho Marx. Dozens of other limousines and Rolls-Royces pulled up for the big event to buy those tickets for $2.75.

Backstage, Bill and his entourage waited in one room, and Steve and his supporters waited in another. Since it was to be a debate, there would be prepared opening statements, but following those, the participants were to think on their feet. Bill Buckley did this better than anyone, and in that sense it was an unfair match. But Steve had prepared as if for war. To guard against freezing up, he had prepared remarks for his rebuttal as well, just in case.

Out front, the crowd was boisterous. This was going to be the event of the century: Steve Allen, head of SANE, the movie-community chapter of the national antinuclear activist group, and Hollywood's favorite liberal, pitted against William F. Buckley Jr., American's leading conservative, who was ready to tell the Soviets that we would nuke them if they made a false move. They were going to march through JFK's foreign policy and examine it from Vietnam, to Cuba, to the Soviet Union. When the debaters took to the stage (Figure 7), the crowd rose to their feet and cheered them into battle. Homer Odum, a local news show host who had also helped me promote the show, took charge as moderator. I walked to the very back of the auditorium in a stunned state. What had I done? There were only two security guards.

Luckily, the rest of the evening took care of itself. Here were

FIGURE 7. More than three thousand of Los Angeles's most politically active citizens from both the left and right came to see the battle of wits between William F. Buckley Jr. and Steve Allen at the Hollywood Palladium.

two great showmen arguing their views. At one point, Buckley spotted Groucho Marx, who was seated in the front row. Sensing that the crowd needed a little jolt, he, without blinking, incorporated the opportunity into his rebuttal. He stared at Steve Allen and exclaimed, "Let's face it, Steve, President Kennedy's foreign policy might as well have been written by the Marx Brothers." Now, most folks hadn't noticed Groucho's presence. He stood up on cue, walked up onstage, and strolled across to thunderous applause, raising and lowering his infamous eyebrows and smoking his cigar all the while.

The blossoming Sol Hurok in me lives on. Over the ensuing years, I am not sure I would have taken on my many professional projects of promoting ideas and debates if I hadn't had this experience under my belt. There is something very intoxicating about taking an empty space and then populating it with vibrant events. Maybe it all helps to ward off ennui. While that turned out to be the only political foray of my life, the dozens upon dozens of scientific meetings I have organized surely grew out of this experience. If done properly, intimate discussions or public debates bring out what people are really thinking. At a minimum, it taught me how the translation of complex topics into the public dialogue worked.

This was the rich and vibrant stew I was living in when all of the science, which makes up the core of this book, was first initiated. Some of the influences came from family, some from the incomparable mystique of Caltech, some from the people of Caltech, some from the people of greater Los Angeles, and some from the incredible good luck of being given an opportunity to study the most fascinating humans on earth.

IN THE FIFTY YEARS since the first studies on Case W.J., which I will describe along with many others, I have studied many neurologic patients with all kinds of illuminating conditions. Of all those patients, this book will focus on the six split-brain patients who have

FIGURE 8. The patients who have devoted so much of their time to our studies over the past fifty years. In the top row (*left to right*) are the founding cases from Caltech: Cases W.J., N.G., and L.B. In the bottom row (*left to right*) are the cases from the East Coast series: P.S., J.W., and V.P.

changed how we think about how the brain carries out its work. These patients are extraordinary in every sense of the word and were not only the center of my scientific life, but a big part of my personal life and the lives of the dozens of fellow scientists who studied them as well (Figure 8). While some have now died, others live on and remain very special people. They are the story and in many ways give the story its very structure. Even with their brains divided for medical reasons, they conquered life with singular purpose and will. How they did this reveals secrets about how those of us without the operation accomplish it as well.

CHAPTER 2

DISCOVERING
A MIND DIVIDED

If I have seen further it is by standing on the shoulders of Giants.

—ISAAC NEWTON

MSG: Fixate on the dot.

W.J.: Do you mean the little piece of paper stuck on the screen?

MSG: Yes, that is a dot. . . . Look right at it.

W.J.: Okay.

*I make sure he is looking straight at the dot and flash him a pic-
ture of a simple object, a square, which is placed to the right of the
dot for exactly 100 milliseconds. By being placed there the image
is directed to his left half brain, his speaking brain. This is the test
that I had designed that had not been given to the Akelaitis series
of patients.*

MSG: What did you see?

W.J.: A box.

MSG: Good, let's do it again. Fixate the dot.

W.J.: Do you mean the little piece of tape?

MSG: Yes, I do. Now fixate.

Again I flash a picture of another square but this time to the left of his fixated point, and this image is transmitted exclusively to his right brain, the half brain that does not speak.* Because of the special surgery W.J. had undergone, his right brain, with its connecting fibers to the left hemisphere severed, could no longer communicate with his left brain. This was the telling moment. Heart pounding, mouth dry, I asked:

MSG: What did you see?

W.J.: Nothing.

MSG: Nothing? You saw nothing?

W.J.: Nothing.

My heart races. I begin to sweat. Have I just seen two brains, that is to say, two minds working separately in one head? One could speak, one couldn't. Was that what was happening?

W.J.: Anything else you want me to do?

MSG: Yes, just a minute.

I quickly find some even more simple slides that project only single small circles onto the screen. Each slide projects one circle but in different places on each trial. What would happen if he were just asked to point to anything he saw?

MSG: Bill, just point to what stuff you see.

W.J.: On the screen?

MSG: Yes and use either hand that seems fit.

W.J.: Okay.

MSG: Fixate the dot.

---

* The location of the brain's speech center in the left hemisphere was discovered by the French physicians Marc Dax and Paul Broca in the nineteenth century.

A circle is flashed to the right of fixation, allowing his left brain to see it. His right hand rises from the table and points to where the circle has been on the screen. We do this for a number of trials where the flashed circle appears on one side of the screen or the other. It doesn't matter. When the circle is to the right of fixation, the right hand, controlled by the left hemisphere, points to it. When the circle is to the left of fixation, it is the left hand, controlled by the right hemisphere, that points to it.* One hand or the other will point to the correct place on the screen. That means that each hemisphere does see a circle when it is in the opposite visual field, and each, separate from the other, could guide the arm/hand it controlled, to make a response. Only the left hemisphere, however, can talk about it. I can barely contain myself. Oh, the sweetness of discovery (Video 1).

Thus begins a line of research that, twenty years later, almost to the day, will be awarded the Nobel Prize.

Take any one time frame from life where many people are involved, and when they retell the story, all participants will have their own version of what went on. I have six children, and Christmas break is a time when the troops all arrive home. Listening to them reminisce about childhood, it is astounding how unique their recall is of the exact same events. The same is true for all of us in our professional lives. While the factual aspects of scientific studies were going on, what was occurring in the background story? Of course, there was more to that magic moment with W.J. than just the two of us.

---

* The brain is a largely symmetrical organ with the left side of the brain controlling the right side of the body and the right side of the brain controlling the left side of the body. The activities of each side of the brain are normally coordinated by the great cerebral commissure called the corpus callosum.

A DARING DOCTOR AND HIS WILLING PATIENT

Bogen was the bright and persuasive young neurosurgeon who pushed along the idea of carrying out the human split-brain procedure (Figure 9). He also was responsible for finding the first case. I could explain how that came to be, but much better are his own words recalling the patient and those early days. From the beginning the revolutionary impact of Case W.J. is evident:

> *I first met Bill Jenkins in the summer of 1960 when he was brought to the ER in status epilepticus;* I was the neurology resident then on call.[†] The heterogeneity as well as the intractability and severity of his multicentric seizure disorder became clearer to me over the next months. Both in the clinic and in the hospital I witnessed psychomotor spells, sudden tonic falls, and unilateral jerking, as well as generalized convulsions. In late 1960, I wrote to Maitland Baldwin, then Chief of Neurosurgery at the NIH [National Institutes of Health] in Bethesda, Maryland. A few months later, Bill was admitted to the NIH epilepsy service where he spent 6 weeks. He was sent home in the spring of 1961, having been informed that there was no treatment, standard or innovative, available for his problem.*
>
> *Bill and his wife Fern were then told of Van Wagenen's[‡] results, mainly with partial sections of the cerebral commissures. I suggested that a complete section might help. Their enthusiasm encouraged me to approach Phil (my chief), because of his experience with removal of callosal arteriovenous malformations. He suggested that we practice a half-dozen times in the morgue. By*

---

* Status epilepticus is a life-threatening persistent generalized convulsive seizure and is a medical emergency. It is traditionally defined as a seizure that lasts longer than five minutes.

† Neurosurgery residents spend time training in neurology as well.

‡ As mentioned earlier, Van Wagenen was the neurosurgeon who first performed callosal sections on humans in the 1940s.

FIGURE 9. Joseph E. Bogen, M.D., was the neurosurgical resident who persuaded his chief of surgery, Peter Vogel, to carry out the first modern-day split-brain surgeries. Joe was a restless intellectual with a great gusto for life and brought a valuable medical perspective to the project.

the end of the summer (during which I was again on the neuro-surgery service), the procedure seemed reasonably in hand. My plea to Sperry was that this was going to be a unique opportunity to test a human with the knowledge from his cat and monkey experiments and that his direction of the research was essential. He pointed out that a student about to graduate from Dartmouth had spent the previous summer in the lab and would be eager to test a human. Mike Gazzaniga started his graduate study in September and was, as Sperry said, eager to test a human subject. He and I soon became friends, and planned together experiments to be done before and after the surgery. There was some delay before the operation, during which Bill underwent testing in Sperry's laboratory. During this delay we also had an opportunity to keep a reasonably complete record of Bill's many seizures.

It was during this period of preoperative testing that Bill said, "You know, even if it doesn't help my seizures, if you learn something it will be more worthwhile than anything I've been able to do for years." He was operated on in February, 1962. It seems to me in retrospect that, if there had been a research committee at our hospital whose multimember approval was required, the procedure would never have been done. At that time, a chief of service could make such a decision alone, which I expect was similar to the situation at the University of Rochester in the late 1930s.[1]

SCIENCE THEN AND NOW

Life was simple back in 1961. Or so it seems now. It was a time when people went off to college, studied hard, went to graduate school, did a thesis, got a postdoctoral fellowship, then got an assistant professorship somewhere. They spent their life pursuing their intellectual interests. Today, the choices are not so clear-cut, and more graduating Ph.D.s go into industry, outreach programs, start-ups, foreign research organizations, and more. Most of one's colleagues are from or have spent time abroad. All of this is fabulous, too, but different and more socially complex.

In the early 1960s, some aspects of biology also appeared to be deceptively simple. Watson and Crick had made their breakthrough discovery about DNA and its role in heredity.[2] By today's standards of molecular mechanisms, the model of how it worked was simple. Genes produced proteins, which then carried out bodily functions. Boom, boom, boom and you had a full mechanism. It became known as the "central dogma." Information flowed in one direction, from DNA out to proteins that then instructed the body. With all that we know today, however, there is serious disagreement on how to even define what a gene is, let alone how many different interactions there are between molecules that are thought to be in some causal chain of action. To complicate matters even more, it is now known that information flows in both directions: what is getting built is, in turn, influencing *how* it is getting built. The molecular aspects of life reflect a complex system laced with feedback loops and multiple interactions—nothing is linear and simple.

Modern brain science started out being discussed in simple linear terms. Neuron A went to neuron B, which then went to neuron C. Information was passed along a path and was somehow gradually transformed from sensory exposure into action, having been shaped by external reinforcements. Today such a simple characterization of how the brain works would be risible. The interactions of the brain's circuitry are as complex as those of the molecules that make it up.

Getting a hold on how it works is almost paralyzing in its difficulty. Good thing we didn't realize this at the time, or no one would have tackled the job.

As I look back on those early days, it may have been good for human split-brain research to begin coming of age in the hands of the simplest of researchers: me. I didn't know anything. I was simply trying to figure it out using my own vocabulary and my own simple logic. That is all I had, along with bundles of energy. Ironically, the same was true for Sperry, the most sophisticated neuroscientist of the era. He had never worked in the human arena and so we held hands as we plowed forward.

In some sense, of course, we all realized split-brain patients were neurologic patients, and neurology was a well-formed field with lots of vocabulary. Joe was our guide in the minefield of jargon. Bedside examination of a patient with a stroke or a degenerative disease was well established and described. The rich history of early neurologists had taught us a great deal about which part of the brain managed what cognitive functions. The nineteenth-century giants in the field, Paul Broca and John Hughlings Jackson, and their twentieth-century counterparts, such as the neurosurgeon Wilder Penfield and still more recently Norman Geschwind, all played major roles in developing the medical perspective on how the brain is organized.

I can still remember the day when Joe came over to Caltech from White Memorial Hospital to give us a lab talk. He described some of our early findings, using the classic terminology of neurology. Although it wasn't gobbledygook, it sounded like that to me, and I remember saying so to Joe and Sperry. Joe was a very open fellow and always progressive. He simply said to me, "Well, go do better," and Sperry nodded in agreement. Over the ensuing years we did, establishing in our first four papers[3] a scientific vocabulary for capturing what was going on in humans who had the two halves of their brains separated.

ORIGIN OF SPLIT-BRAIN RESEARCH

Split-brain research in animals has a rich history. This all occurred before my time in the lab, and it is easy to imagine that there are many versions of the story. The most straightforward begins with Ronald Myers working on a M.D./Ph.D. degree at the University of Chicago in the mid-1950s. His project was to learn how to cut the optic chiasm down the midline in a cat—a formidable assignment. The chiasm was seemingly inaccessible. Located at the base of the brain, it was where some of the nerves from the left eye and right eye cross, allowing information from both eyes to project to each half brain. If he could successfully cut the chiasm, it would mean visual information coursing up from the right eye would stay lateralized— that is, it would go only to the right half brain—and information coursing up from the left eye would go only to the left brain. The surgery would have eliminated the normal information mixing at the base of the brain.

If such a surgery could be done, then it would mean that one could begin to test how information from one eye came together inside the brain with information from the other. All of this was driven by the working hypothesis, then unproven, that the neural structure integrating the information was the corpus callosum, the huge nerve tract that interconnects the two half brains. There were those, such as Karl Lashley, mentioned earlier, who thought that the corpus callosum was merely a structural element that supported the two hemispheres. The experiment Myers designed was meant first to teach a visual problem to one eye of a chiasm-sectioned cat and then to test the other eye. If the information was integrated, then the idea was to test again after cutting the callosum to see if the integration stopped. The prediction was that it would. That would be huge.

Myers worked on the procedure and finally perfected what was, at first, an extraordinarily difficult technique. After much practice it became quite straightforward even though it doesn't sound at all easy. His original description is telling:

> The optic chiasma was transected in the mid-saggital plane through a transbuccal [through the mouth] approach. In this procedure the soft palate was incised from its attachment to the hard palate anteriorly to within a half-centimeter of its free margin posteriorly. The cut edges were retracted with catgut sutures creating a diamond-shaped opening. A flap of nasal mucosa was next reflected from the sphenoid bone, and, with a dental burr, an oval fenestra 1 by 5 mm. was made in the bone immediately anterior to the spheno-presphenoidal suture. Through this opening in the bone the dura was carefully exposed and incised, thus revealing the underlying optic chiasma. The chiasma was then sectioned with a fine steel blade, under close visual control through a binocular dissecting microscope. A small piece of tantalum foil was inserted between the cut halves of the chiasma so that post mortem verification of the completeness of section would be possible by gross inspection.
>
> After section of the chiasma, the opening in the bone was filled with Gelfoam soaked in blood to form a barrier between the nasopharynx and cranial cavity. The flap of mucosa was replaced over the Gelfoam and the incised soft palate reapposed with catgut sutures.[4]

Got it? Myers was set to perform his experiment. He found that in the chiasm-sectioned cat the information was integrated, and just as he predicted, after the callosum was sectioned, the integration stopped. This procedure, along with the finding that the corpus callosum transferred information between the two hemispheres, launched a thousand ships. With both surgeries, now each hemisphere could be directly given visual information and the opposite hemisphere could be tested for its knowledge of the information.

With Myers's chiasm surgery breakthrough in hand and the logical next step to cut the callosum, interest was developing in what first seemed like a relatively obscure, yet confounding finding. Akelaitis's patients at the University of Rochester appeared to have no major behavioral or cognitive changes following callosum surgery. As a consequence of this work and Lashley's stance, most people thought that when it came to humans, little would come of this careful new animal work of Myers and Sperry.

Of course, one of the beauties of science is that it marches on. As the split-brain story developed and became rich and influential in science, people wanted to know where the idea came from. Was it Myers? Sperry? Both? Others? Did it just slowly happen as information accrued over time? After all, it wasn't until years after Myers did his work that the whole preparation was dubbed "split-brain" by Sperry,[5] the consummate wordsmith.

One account of its origins came from a well-known psychologist, Clifford T. Morgan, who had moved from Wisconsin to Santa Barbara in the early sixties. He had been an instructor at Harvard in the early forties and no doubt had known Sperry, as they both were associated with Lashley. Morgan was keenly interested in epilepsy and also became a celebrated textbook writer. His first book, *Physiological Psychology,* published in 1943, was credited for bringing order to the field by systematizing its many facets.[6] Morgan went on to a distinguished career, started his own publishing company, his own journals, and his own society. Perhaps he was the model for my own subsequent entrepreneurial efforts to start a journal and a scientific society.

I later met Morgan at his office when I arrived for my first stint at the University of California, Santa Barbara (UCSB) in 1966. He was a warm and generous man who seemed to live in order to hear Dixieland jazz at a local spot, the Timbers, on Sunday nights. In fact, he was so generous that on the spur of the moment one day, he lent me five thousand dollars to help me buy my first house! Just like that he wrote out a check at his desk with the simple command, "Pay it back when you can," and handed it to me. That simple gesture kick-started my domestic life and had a big impact on me. Years later, following his example, I was able to do the same for two of my young research associates.

It turns out that the idea for the split brain was spelled out in the second edition of his book in 1950, coauthored with University of Pennsylvania psychologist Eliot Stellar.[7] It was stated with no fanfare and made to sound as if it were part of the culture at the time, whereas in fact everybody wondered what the callosum did, and everybody wondered how information was communicated between the hemispheres. Does it remind you of what was going on in the field of genetics? After all, everybody knew there was inheritance and everybody knew there was DNA before Watson and Crick put it together. Maybe major advances simply accrue. At the same time, and very importantly in my view, somebody has to go out and do something to prove or disprove the talk, not just go on and on about it. There was no question in my mind Myers and Sperry had gotten their hands dirty and transformed the talk into findings.

I met Myers years later at a conference where I was presenting the human split-brain work and he was presenting some of his anatomical work carried out on chimpanzees.[8] I was eager to get to know him because I fully understood his crucial role in the history and development of split-brain research. As a scientist he certainly had earned the respect of his peers, and the field of brain science was indebted to him.

That doesn't mean he was Mr. Nice Guy. After my talk, he went into some kind of rant about how the "odd human case" didn't mean much of anything and that it was sort of a bizarre consequence of prior epilepsy, etcetera. I was stunned and rather speechless. But the lightbulb slowly was turning on. Turf is king, and I was on his turf, even though I was following up his work in another species, and, by that time, our human studies had been peer reviewed in several refereed journals. I was getting another lesson in the difference between scientists and science. I was also wondering if it was inevitable that all contributors to intellectual property turned out this way. Was there any difference between an artist, a scientist, a bricklayer? Would I also turn out that way? Note to self . . .

DR. SPERRY

Roger Sperry was a true giant in the field. When I arrived at Caltech, he had recently recovered from a relapse of tuberculosis. His wife, Norma, coordinated the flow of information to him from the lab, while he rested and recovered at the sanatorium. At that time, he was involved in at least three major scientific projects. His foundational work in neurobiology, which was revealing that animals were not randomly wired and then shaped by experience,[9] was going strong. He also proposed the bold hypothesis that a chemoaffinity process was in play—a process that guided neurons to grow to a specific destination during development. He had outlined that idea at a conference a few years before, and it served as the basis for Caltech hiring him into a professorship.

Sperry had taken on another issue. There was something called psychophysical isomorphism.[10] This was the idea that, for example, if one saw a "triangle" in the real world, there was a corresponding electrical pattern in the visual brain areas that matched the real-world picture. To test this idea, he inserted little mica plates into the cortex of cats. The mica served as an insulator so that any electrical field potential in the brain, should it exist, would be highly disrupted by the many intervening insulators, thereby preventing the animal from performing a visual perceptual task. Many variations of this experiment were carried out. All the results supported Sperry's belief that the notion of psychophysical isomorphism (parallelism) should be abandoned. It has been.

On top of all this, of course, was the exploding research on split-brain animals. Sperry had an army of postdoctoral research assistants working mostly on cats and monkeys. The lab was going full tilt on a variety of issues that dealt mainly with the question, Would an animal with its corpus callosum sectioned show transfer of information between the two hemispheres when a perceptual problem was trained to only one hemisphere?

Any one of these thrusts of research would have been enough to keep most labs busy and noticed in the larger scientific community. Sperry had a style that let things happen. He didn't tell us to how to do science. He watched, he kibitzed, he surely guided in ways we didn't fully understand at the time. When he saw something of interest, he knew how to bring it out and enhance it. Put differently, he had a nose for the important versus the routine.

More generally, those of us who have spent a life in science running large labs wonder how it all keeps going. It most assuredly does not happen by the lab director issuing new directives on a daily basis. Labs can go for years with yeoman-grade science taking place. There can be dry periods, dull periods, nonfunded periods. Occasionally, however, something—sometimes it's serendipity, sometimes it's an actual hypothesized experiment—comes along and works out. Instantly, all the mundane days dissolve into glee and excitement.

I can remember George Miller, the distinguished psychologist, saying to me, "Everybody wants to think science moves forward one clean hypothesis at a time. It moves forward, but usually by stumbling on to something that was unintended." Yet we then quickly tell a story about how we logically proceeded to our findings, which keeps the myth going. Science is great, but scientists are human and prone to storytelling just like everyone else.

Nonetheless, keeping an overall lab narrative going is crucial to keeping the research focused and on track. Young scientists come and go. They make their contribution to part of the story, and in return, they receive the support of the lab director throughout their career. That is the standard arrangement. Students usually continue working on an aspect of the problem they contributed to, and in the long run, that is how major research thrusts develop. Even pedestrian lines of research can grow like this.

The successful labs keep an edge by having really smart students and postdocs. Of course, smarts aren't the only ingredient to success. Everybody is smart, but some students are also energetic and practical. A further combination of hard-to-predict characteristics and luck—which is what I had when I hopped into this lab's ongoing dynamics—lead to a successful career in science.

Back in my undergraduate summer at Caltech, my meeting with Sperry in his Kerckhoff Hall office was the first of many experiences of meeting "the man." His scientific reputation was, as I said, exceptional. From neurodevelopment to animal psychobiology, he was the intellectual leader of his time.

People really do have two realities—the everyday person and the "metro" person, or the private self and the public self, as its commonly phrased. The public self is your job, your reputation, the model the world builds about you and expects of you. It is usually not you. Let's face it: If Keith Richards lived the life we all think he lives, he would be dead.

We can sometimes come to be ruled by the metro self. We live to feed it and do what it tells us to do. This thing that isn't the real you is now running your life, making demands on you. Meanwhile, the real you is trying to get the kids to school, root the gophers out of the rose bed, see your friends for a drink, and talk about whatever. In my life, twenty years of lunches with Leon Festinger, the distinguished social psychologist, demonstrated that someone who had a large metro self could also be exceptionally personal and not let that self intrude or take over his private life. Many people pull this off.

I have always been amused by my many colleagues who claimed they knew Roger Sperry. They knew the metro Sperry. I can say with a fair degree of confidence that nobody knew him like I knew him, both his everyday self and his—legendary—metro side.

DISCOVERY AND CREDIT

With that glorious day testing Case W.J., and revealing that sectioning the callosum in humans had an effect in line with the preceding animal work, the fifty-year program of study on human split-brain patients began. It was luck that I was there. Sperry let me flourish, as did Bogen. Others in the lab, who were interested in the results, let me remain in charge of the project. It was a time of good fortune.

Our first report was a brief communication to the *Proceedings of the National Academy of Sciences.* Sperry had recently been elected to the academy, and members in those days had a fast-track way of publishing. We worked like crazy through the winter and spring and got the paper off in August 1962 for an October publishing date. The paper, largely free of medical jargon, was an amazing case history, a succinct summary of all we had done on W.J.[11] The idea that disconnecting the hemispheres of the human brain caused major effects had new life. The era of human split-brain research was born.

At the same time, another story was brewing, one that would begin to teach me about the competitive nature of scientists. Norman Geschwind (Figure 10), a young neurologist, and Edith Kaplan, an equally young neuropsychologist, were working at Boston Veterans Administration Hospital. They reported a case of a

FIGURE 10. Norman Geschwind played an early role in establishing the idea of having neural disconnection syndromes in neurology. He is generally credited as the father of behavioral neurology in America. Being in his company was always a pleasure.

patient, P.K., who suffered from a gliablastoma multiforme, a tumor that had invaded his left hemisphere. During surgery, presumably to debulk the tumor, he sustained an infarction* of the anterior cerebral artery.[12] As Antonio Damasio recounted years later in Geschwind's obituary, "The anterior section of the callosum as well as the medial aspect of the right frontal lobe were destroyed" and "resulted in a severe disturbance of writing, naming and praxic control of his left hand."[13] In short, a natural lesion resulting from a stroke, as opposed to a surgical section of the corpus callosum, had revealed a disconnection effect.† They first reported their finding at a December 14, 1961, meeting at the Boston Society of Neurology and Psychiatry.[14] Geschwind and Kaplan had very cleverly interpreted a messy tumor case as a callosum lesion case and carried out some simple tests that suggested they were right. After the patient died a few months later, their diagnosis was confirmed at autopsy. In the spring, another posting of the tumor case was made. In the newsy Random Reports section of the May 1962 issue of the *New England Journal of Medicine* was an entry made by Geschwind about these astounding observations. Great care was taken to note that the report came out of the December 14, 1961, meeting. The finding was the buzz in Boston.

Geschwind sent a copy of his prepublication manuscript to Sperry for comment sometime during the early months of 1962, right when we were testing W.J. but before we had published our findings. In the manuscript, he credited Sperry's animal work for his and Kaplan's idea of examining their patient for disconnection effects. Sperry had given a colloquium at Harvard during the fall of 1961 describing some of that work. The broad outlines were already known by then, and it was entirely routine for a researcher to talk

---

* A cerebral infarction, or stroke, occurs when an artery that supplies a part of the brain becomes blocked or leaks. The area of tissue that loses its blood supply dies.

† A disconnection effect is a neurological disorder caused by the interruption in the transmission of an impulse along a cerebral nerve fiber/pathway.

about their "latest" findings. At that time, I was working feverishly back in Pasadena testing W.J. preoperatively. Sperry was not pleased to receive the manuscript a few months later. Both research groups were working entirely independently and he didn't want any potential confusion on that point. After all the years of his and Myers's foundational animal work and our early and, as yet, unpublished work on W.J., he didn't want anyone thinking that his human work stemmed from Geschwind's findings.

Roger Sperry was a fierce competitor—an athlete who had lettered in three varsity sports in college. During the early part of my training, Sperry got into a huge feud with his own mentor, Paul Weiss, then the most eminent neurobiologist there was. In a heated, specially arranged addendum to a Weiss summary in something called the *Neuroscience Research Program*, Sperry let him have it. In an autobiographical essay about her illustrious career, another of Weiss's students, Bernice Grafstein, recounted it like this:

> *I was greatly relieved, therefore, that my failure to progress with the problem of motor system regeneration, when I eventually summarized my results for Weiss toward the end of 1964, did not seem to trouble him particularly. He seemed not at all perturbed that my findings might be less consistent with his ideas than with ideas of specific reconnection that were identified with Roger Sperry. In fact, in his summary report of a workshop session at the Neurosciences Research Program at about that time, Weiss claimed to embrace the idea of specificity in regeneration, with reservations only about the necessity of uncovering the detailed mechanisms involved [although he still insisted that these might include functionally coded activity patterns that could serve as "messages for selective reception].*[15]
>
> *Sperry, on the other hand, was adamant about disengaging his views from any associated with Weiss. In a statement that he insisted on appending to the same report, Sperry reasserted his own primacy in the development of the idea of "selective, chemotatic*

(sic) growth of specific fiber pathways and connections governed by an orderly pattern of specific - chemical affinities that arise out of . . . embryonic differentiation" (Sperry, 1965). He believed that throughout their long association Weiss had assimilated his (Sperry's) contributions without adequate acknowledgment, and that there had been "a buildup in the literature of a complex web of ambiguity, forced terminology, and confusion of issues that [was] almost impossible to untangle for anyone not intimately acquainted with the underlying history." He was not content that Weiss should just confirm that specificity was operating in the growth and termination of regenerating axons; he believed that he had been deprived of the opportunity that Weiss had promised him to publicly "get things out in the open, face the issues and clarify points of controversy."

Clearly throughout his career you were either on Sperry's team or the other guy's team. Geschwind was a new rival. It was also true that if someone graduated from Sperry's team and became a competitor, the frequency of critical remarks about that person went up. During my stint at Caltech it happened every time. One day when he was criticizing somebody who had left the lab, I realized that after I earned my degree and left the lab this would probably also happen to me. At the time, however, we were on the same team and I shrugged the thought aside, with assumption that it was just part of life in science.

Geschwind's complete findings, ultimately published in the journal *Neurology* in October 1962, served an important role in activating the interest of neurologists in the callosum. It reconnected the clinical literature with a rich earlier history about the importance of the callosum, previously worked out by the German neurologist Hugo Liepmann and the French neurologist Joseph Dejerine around the turn of the twentieth century.

Years later, Geschwind and I were guests at an International

Neurology Meeting in Kyoto. It was a very formal event at a large conference hall. The auditorium was lined with translators madly trying to deal with the multiple languages being used. Norman was on the dais with a group of famous neurologists from around the world. The emperor of Japan was also on the dais with his wife, listening to what must have seemed like gibberish. Each speaker always rose and bowed toward the emperor before going to the podium to speak. Not Norman. When he was called up to speak, he went directly to the podium, said his piece, returned directly to his chair, and sat down.

After the session I asked Norman about it. His failure to conform was noticeable, to say the least. Norman said, "Hell no, I am not going to bow to the emperor of Japan after what he did to our troops." He was indeed an honorable man and a competitor. Over the years, I became a friend of Geschwind's and friends with the entire Boston VA's neuropsychology group. If we ever spoke about the issue of precedence, I don't remember it, and we had dozens of opportunities to do so. Norman was a scholar and a conversationalist like few others. It was always a joy to be in his company. He wrote a paper in 1965 for the journal *Brain* that to this day is a classic review of the neurologic "disconnexion" syndromes (as it was spelled in the British publication).[16] Indeed, the paper launched the field of behavioral neurology in America.

While Geschwind's unsolicited manuscript was passed around Caltech, it didn't have much of an impact on our thinking. Sperry said that whenever someone makes a discovery in science, someone else always says, "Yeah, but so-and-so thought of it before you."

In many ways, Sperry was more socially conscious than most. He was always thinking about how his actions might impact the social fabric of scientists. Bogen, in his autobiography, tells of another story that captures this quality. He was discussing an exception to Sperry's leisurely way of getting a manuscript out for publication:

Roger did not always delay. One day when I was visiting the lab I asked him about the Gordon paper on lateralized olfaction in split-brain patients. He said, "We have to send this olfactory paper in immediately." "Why?" I asked. "Because I have just refereed for Neuropsychologia *a paper with a similar experiment in rats. People know that with human subjects, we can do in a few weeks what would take many months in rats. If we delay, people might think that I got the idea when refereeing the rat paper." Roger seemed to think of everything. I idolized him and hung on his every word, of which there were not very many. I thought him the experimental physiologist of our time.*[17]

At that time, I was way too callow to comprehend the complexities of sharing credit for an intellectual idea or that it's a constant battle to take the scientist out of science. Unfortunately, it is now commonplace to have authors suggest to journal editors a list of preferred *and* a list of nonpreferred reviewers. This recent trend has arisen because many people have realized that pettiness has held back many scientific developments. New ideas need a chance to be expressed. This practice, however, under the rationale that it would be a "conflict of interest," also disallows people from critically interacting. Should someone really be disqualified to review a paper because they have a different interpretation of the underlying data? That is anathema to the very nature of science.

At the time, I just kept doing experiments, and after a while the whole manuscript exchange episode passed. After all, we had already realized that the disconnection story and the loss of some capacities was not the most profound implication of the split-brain studies. We had begun to understand that we could test each half brain separately, independent of the influence of the other half. Unlike classic neurology, where you study the absence of mental capacity caused by lesions in particular areas, we could study the *presence* of mental capacities. It was a whole new ball game.

ESTABLISHING THE BASICS

Though the trembling excitement of the discovery would soon pass, we knew we had a research gold mine on our hands that could explain some of the brain's mysteries. The slow, careful exploration of what we had to do to confirm and extend the basic findings needed to begin. Right off the bat, we ran into a complicated problem. Our original paper in *PNAS* had been mostly about limiting visual information to one hemisphere or the other. This was relatively easy to do. The next phase called for limiting touch information to one side of the brain. This was not easy at all.

The visual system in humans and similar mammals is neatly laid out in our body plans. Stare ahead and look at one spot. Both of your eyes are pouring visual information into your brain. Does it enter in an orderly manner? Yes, it does. Each eye sends its information up the optic nerve and half of the information stays on the same side of the brain and half of it crosses over and goes to the opposite hemisphere (Figure 11). So, if you are still fixating on that point, everything to the left of the fixated point in either eye is projected solely to your right hemisphere. Thus each eye is contributing to that experience. It follows, then, that visual information to the right of the fix-

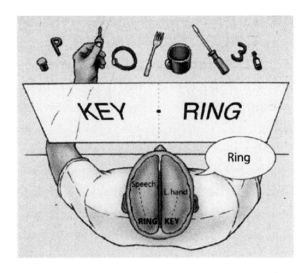

FIGURE 11.
Schematic of how information is projected into the brain by the visual system.

ated point is being solely projected to the left hemisphere. This is true for all of us, including our split-brain patients.

This makes it easy to test each hemisphere separately when using visual stimuli. One simply has to present whatever it is you are interested in knowing more about to the right or left visual field. Once again, information from the right visual field goes to the left hemisphere and information from the left visual field goes to the right hemisphere. Got it? Then you are ready to start thinking about these experiments.

Working out the strategy for testing how a separated hemisphere would deal with touch—or, more formally, somatosensory information—is more challenging. How the brain receives information from the body is quite different. This was beautifully laid out by Jerzy Rose and Vernon Mountcastle in a chapter of the 1959 *Handbook of Physiology*,[18] which I read at the time. They were the world's authorities and their clarity was inspirational.

Here is how it works. The left half of your body sends most, but not all, of the information about touch to the right hemisphere. If you are holding an object in the left hand, the touch information related to the object's overall shape, called stereognostic information, goes to your right brain. More basic sensations associated with the mere presence or absence of having been touched, however, go to both hemispheres. Rose and Mountcastle made this abundantly clear by also describing the anatomy that supports this reality. The reverse is true for the right half of the body. Information from the right hand about an object's shape goes directly to the left hemisphere, while the less definitive presence or absence of information goes to both hemispheres.

Clearly, from the perspective of getting completely lateralized information into only one hemisphere, the visual system was the way to go: simple, clean, and highly lateralized. The somatosensory system, however, was presenting a challenge. Some forms of the information from the world of touch went to the opposite half brain while others went to both half brains. How were we going to make sense of this? It turned out to be intriguing, thanks in large part to the work that had gone on before us.

To solve this puzzle, we first blindfolded the split-brain patient and placed an object in the right hand. Then we asked, "What do you have in your hand?" The object was always named correctly: no fuss, no muss. The shape information had gone to the left hemisphere. Then we placed the object in the left hand and asked the same question. This time the shape information went to the right, nonspeaking hemisphere. The patients were usually not able to name it. Interestingly, however, they would manipulate the object appropriately. This suggested that their right hemisphere "knew" what the object was but because it had no speech center, it couldn't name the object. Nor could the shape knowledge of the object be communicated to the left speaking hemisphere. The fact that the object was manipulated correctly also indicated that both hemispheres had stored information about the nature of objects, sort of a double memory system leading to redundancy in our brain organization. All of this from one bedside test:[19] Fantastic!

One sunny afternoon I was testing W.J in his home in Downey. I can still remember how much delight he showed on the following test. I had prepared a set of small wooden blocks that had small tacks protruding from them. I was looking to see if he could tell the difference between a block with one tack and a block with several tacks. I blindfolded him and started presenting the blocks, first to the right hand, which found it easy to do, and then to the left hand. The task was simply to match the blocks. I would first give one of them to him, then take it away and put it in with a group of blocks. He would then pat around on the tabletop and try to find it again among the other blocks. It turned out each hand could do this simple "match to sample" task.

What was most interesting, however, was what his left hand (under control of the right brain) would do when presented with the block with one tack. He would pick it up by the tack and twirl it. It seemed as if his right brain was showing off his dexterity with the hand it controlled, even though it could not relay information about what the object actually was. He also was chuckling to himself while he was doing this. It seemed as if his right hemisphere was an independent personality enjoying the moment. It was one of my first realizations in these early days that there were "two minds" present at all times. I remember asking W.J., "Why are you laughing?" He replied, "I don't know. Something in my left hand I guess."

What was puzzling, however, was the fact that sometimes W.J. did name an object held in the left hand correctly. How was that happening? How could that work? It took months before I finally figured out what now appears to be an obvious answer. As Sperry use to say, "Nothing is simpler than yesterday's solutions."

The key was to remember those neural pathways and the double representation that Rose and Mountcastle had written about: Some of the fibers from the somatosensory system do not cross over to the opposite half brain. They climb up ipsilaterally, that is, to the half brain along the same side as the point of stimulation. It was unclear, however, what these fibers were doing. Finally, I hit upon the experiment that revealed the answer.

I limited the number of objects to be identified to two, a plastic triangle or a plastic ball. All W.J. had to do while blindfolded was say which one I had placed in his left hand. After a few trials, W.J began guessing correctly on every test trial. How was he doing it?

Imagine you are given such a task but are required to wear a thick leather garden glove. The instant recognition of the nature of the object would be gone due to the muffled information you would be getting through the glove and you would have no immediate stereognostic information. How could you figure it out? You would quickly learn to find an edge on the object and press hard on it. Presence or absence of information is the kind of minimal signal those ipsilateral pathways can transmit. In a split-brain patient, without any shape information coming in from the right hemisphere, the left (speaking) brain would quickly learn "I feel an edge" or "I don't feel an edge" and then conclude: If an edge, it must be a triangle; if I don't feel anything, then it must be a ball.

That was exactly what W.J. was doing. He would manipulate the object until his thumb could press down hard on an edge. That became the cue that initiated the whole cascade of events I just described. This was one of the first realizations that split-brain patients use external self-cueing to reintegrate some of their disconnected information. By self-cueing I mean a behavior that is induced by one hemisphere and perceived (through one or more of the senses) by the other hemisphere. This in turn allows that other hemisphere to initiate an appropriate response. It is startling when you first see it.

This seemingly simple observation points up a deep problem for those of us trying to figure out how the brain works. As will become more apparent later on, many researchers consider the brain to be made up of dozens if not thousands of modules. A module is a local, specialized, neuron network that can perform unique functions and can adapt to or evolve to external demands. Modules work independently, yet in some kind of coordinated way, to produce unitary behavior. Think of a city with hundreds of different independent businesses. Taken together, however, they perform all that needs to get done for a city to operate and appear to be a unified whole. How all the modules are coordinated is the question. Over the years, it has become apparent to me that one way the modules come to produce unity is by cueing each other, usually outside the realm of conscious awareness. The self-cueing seen in a disconnected speaking hemisphere is omnipresent, and we have observed many different strategies. In this case, primitive touch cues are picked up and then that information is intertwined with the limited decision set of two possible items to render a correct answer. This is but one strategy. But I've gotten ahead of myself.

FIRST FILMS:
THE BEGINNINGS OF THE LEFT BRAIN/
RIGHT BRAIN DISTINCTION

Ideas for things we should study kept rolling in. In the beginning we took them on, or at least initiated them, quickly and rather easily. While Sperry was unquestionably interested in the studies, they had not yet captured his full attention. As I already mentioned, he was extremely busy with his other exceptional projects. Like a financier, he had placed a bet on a small company, of which I was the chief operating officer, and he was waiting for enduring returns. He kept closely involved with almost daily meetings with me, but he had a way of keeping a distance as well. As Bogen wrote in his autobiography:

> *Then a lot of things came after that. In the beginning Sperry was not that interested. He just thought he would let me and Gazzaniga do it. But it became apparent to Sperry after the second patient that anything you could do with a monkey you could do a lot faster with human beings. He got a lot more interested.*[20]

Back east, Norm Geschwind complained a bit about his original observations not receiving instant recognition beyond the local Boston buzz. He wrote: "As a reflection of this lack of interest it is interesting to observe that when Edith Kaplan and I described the first modern patient with the syndrome of the corpus callosum, the paper was rejected by the *New England Journal of Medicine* without any comment."[21]

The journal *Neurology* was interested, however, and accepted it. It is not unusual for big discoveries not to be recognized as such. Paul Lauterbur's first attempt at publishing his work that led to the development of MRI (magnetic resonance image) was rejected by the journal *Nature*. He went on to win the 2003 Nobel Prize in Physiology or Medicine for it. He later quipped, "You could write the entire history of science in the last 50 years in terms of papers rejected by *Science* or *Nature*."[22]

I have come to realize that, for a short while, the enthusiasms of discovery can be quite local to the person doing the work. The world is a busy place and everyone is vested in what they are doing. Grabbing their attention away from their current passions is a process. David Premack, one of world's most talented psychologists, once observed that after he published his paper in *Science* (groundbreaking studies that led up to his theory of motivation—what later became called the "Premack Principle"), he thought it would all be easy from there. He remarked to me, "Little did I know I would spend the next ten years hitting the road to sell it at every departmental colloquium I could find."

Sperry's interest did indeed come around. I would be left with no doubt about that after a talk I was asked to give to the biology department, a sort of sounding-board seminar for graduate students. When I gave the talk, I had been working on the split-brain project for at least two years and was deeply into all aspects of the testing program. But it was the films I showed that were the showstoppers.

From the beginning, I had been intent on establishing a film record of the experiments. In 1962, video cameras did not exist. It was all film, 16 mm film to be exact, with clunky cameras. That found me making friends with the people at Alvin's camera shop, on Lake Street in Pasadena.

Filming is not an easy enterprise. At the beginning, I didn't know anything about filming: about lighting, f-stops, focus, depth of field, or keeping the object to be filmed in view. Most of all, I didn't have any idea how much a Bolex camera cost! The people at Alvin's camera shop taught me everything. They began by fixing me up with a hand-cranked Bolex 16 and a tripod. I started out easily enough. One of the questions that kept coming up whenever anybody heard about the work was, How were the patients affected in everyday life? So off I went to film patients doing everyday activities. One of the patients, N.G., lived in West Covina (about thirty minutes from Caltech), and her husband worked at a Ford factory. They lived quite well and had a lovely swimming pool in their backyard. One day I went out to see her and set up the camera by the pool. N.G. gladly swam a couple of laps. I then set up a shot of her sitting down on her couch and reading a newspaper, just like anybody else (Video 2).

To get more sophisticated than this required another addition to the camera. You could attach a small motor that allowed you to "film" remotely. Thus, I could set the camera up on a tripod, direct it toward an aspect of the experiment, and control the camera while I was doing whatever had to be done in the experiment. These films proved quite dramatic.

And of poor quality: I was definitely not an ace cameraman. What to do? My good fortune was that I had come to know Baron Wolman, a fantastic young photographer and a future founder of *Rolling Stone*. He somehow had become totally fascinated with the research (another in the growing list of nonscientists I was meeting who were interested in science) and was willing to lend me a hand. I asked if he would mind driving down to Downey with me to film W.J. doing a simple task on a tabletop. Baron didn't mind and came along. That film has become one of the benchmark representations of split-brain research (Video 3).

The film was clear and crisp. In it, W.J. was first requested to arrange a set of blocks that were part of the Wechsler Adult Intelligence Scale test bank. These were four blocks each colored differently on their six sides. The patient was shown a card with one of many possible patterns on it and was asked to simply arrange the blocks to match the pattern, commonly called the Kohs blocks. For reasons that are still not completely clear, the human right cerebral hemisphere is specialized for the kind of visuomotor function that enables someone to do this task. This would predict that the left hand, which gains its motor control from the right hemisphere, would be superior at doing this task. That is exactly what happened and is shown in the film. Lickety-split, the left hand assembles the blocks directly.

The next scene shows the right hand trying to do the same task. The right hand gains its control from the left hemisphere, the one with language and speech. Yet, when it tried to assemble the four blocks to match the very same picture, it couldn't do it. It couldn't even get the overall organization of how the blocks should be positioned, in a 2x2 square. It just as often would arrange them in a 3+1 shape. It was amazing. And as the right hand was trying and failing, suddenly the more competent left hand would try to intervene! This was such a common occurrence that we had to have W.J. sit on his interfering left hand when the right hand was trying to do something.

Finally, we ran a trial where we left both hands free to try to solve the problem. In that one sequence, a much larger picture of what split-brain research was ultimately going to teach us was revealed. In brief, one hand tried to undo the accomplishments of the other. The left hand would make a move to get things correct and the right hand would undo the gain. It looked like two separate mental systems were struggling for their view of the world. It was dramatic, to say the least.

The entire Kohs block test grew out of two earlier efforts. First, there was clear evidence from earlier neurological studies that right hemisphere lesions impaired the ability to draw a picture with a three-dimensional perspective, such as drawing a cube. We first did this test very early on, and it was clearly the case with W.J. The left hand could draw a respectable cube. The right hand could not.

Meanwhile, Bogen kept trying to convince us to do "standard neuropsychological testing." Sperry and I didn't see the point, but to his credit, Joe pursued the matter and writes about it convincingly:

After Bill had recovered from his surgical ordeal (and was feeling better), he was eager to participate in some laboratory experiments. After some months a helpful social worker got in touch with a psychologist who occasionally tested clinic patients. She arranged some funding and he agreed to meet. He seemed to me not only quite elderly but actually quite infirm.

I explained the patient to the psychologist and how interesting he was. I asked him, "Do you give the standard tests?" "Oh yes, the Wechsler." I didn't know much about the test, and neither did Mike, and he reluctantly agreed after some argumentation.

"Old Daddy Edwards," as I learned he was sometimes called around the hospital, acceded to our request. He sat across a card table (part of his equipment) from Bill. Mike and I sat on the other two sides watching. The testing went along for an hour or so, somewhat tediously from our point of view, until Dr. Edwards pulled out the Block Design subtest. Bill pushed the blocks around somewhat ineffectually. Meanwhile Edwards was timing in his usual fashion and ended up with a zero score. I suggested that [Bill] use one hand at a time. Dr. Edwards objected because it was customary for subjects to use both hands. However, he was persuaded to try this momentarily, so we asked Bill to use only his right hand . . . while sitting on his left hand. . . . He had considerable success. Mike and I looked at each other as if we had caught a glimpse of the Holy Grail. "Now try it with just your left hand," I asked. He was quite successful! "Now try the next pattern." With his left hand he did the next one quite quickly. "No!" Edwards said. "He is supposed use both hands." It was getting a little tense, because he insisted on doing it the standard way and we were anxious to further pursue our Grail. Dr. Edwards quietly prevailed, and he finished the tests. We thanked him, and he replied, "Yes, it was interesting. We should test 20 or 30 more of these epileptic patients with various lesions." So this fumbling with the two hands seemed to be an example of what Akelaitis called "diagnostic dyspraxia," which we had subsequently termed "intermanual conflict."

We realized Edwards was in the dark as to what had happened and what sort of patient Bill was, or why we were so wreathed in smiles. My next move was to borrow a set of these Wechsler blocks (one needed a license to buy them), and then I eventually obtained a set of the Kohs colored blocks. We retested Bill, and sure enough he had the same discrepancy between the

left hand doing well, and the right hand doing poorly, and this
persisted for at least 2 years. When we showed the data to Sperry
he commented in his usual soft, skeptical way, "Well, I guess you
boys have got that fellow pretty well trained by now." It was true
that not all patients showed this discrepancy. The second patient
did not show this discrepancy, being actually rather poor at the
test with either hand. However, some patients definitely did show
the discrepancy.[23]

Thankfully, in the late 1960s, soon after my Caltech days, the black-and-white reel-to-reel video cameras arrived. Combined with the recording unit, they were originally large and clunky, even though they enabled the video recording of experiments. Still, at least in the early days of video, something about "film" had video beat by a long shot. A few years later, when I moved to the State University of New York, Stony Brook, my entire start-up package went to purchasing a Beaulieu News 16, a 16 mm camera that recorded sound on the film. The silent movie era was over. We could actually hear what patients said in response to questions and queries of various sorts. I hauled that camera around in its fancy huge aluminum case for years, taking it to Paris and other places, feeling quite macho while doing so. Many years later it had become a burden to maintain and use, and it slipped off the inventory. A talented Dartmouth undergraduate was graduating and leaving for a life in documentary filmmaking in Brooklyn. I had it tuned up and gave it to him. It lives somewhere.

As I have said, Sperry was about to become fully engaged. At my coming-out seminar, I had put together a fifteen-minute film of the patients doing their various things. First I showed them in everyday-life situations looking utterly normal in all ways. Next were various scenes of patients showing the disconnection effects, including not being able to speak about visual information projected into the left visual field. More important, I then showed scenes where that left-field information could nonetheless be acted upon with the left hand by finding a matching object from a group of objects. Finally, the film wound up with the segment of W.J. doing the block test I described. It was, to say the least, riveting: a showstopper. Everybody said so. That was a great day.

I would describe the next day differently. It was tough. Sperry asked me to come to his office, and within minutes the grilling started. He challenged every finding as if he had not been following them closely, even though we had spoken for hours upon hours after each testing session, many of which he had attended. He went at it for a couple of hours. I was in shock. I didn't realize for some time that he was doing the right thing. The fifteen-minute film really made the studies come to life, and he knew this was going to be a big deal. He wanted to make absolutely sure all the tests were done correctly. He had placed his bet and given his complete and unwavering support early on. Now he was going to really make sure the work was unassailable. He was doing his job, and I was still learning how it all worked.

In the background was the ever-present Caltech culture, a culture that was all about inquiry. Richard Feynman, for example, had a way of popping into graduate students' offices and asking them what they were doing. One day I was working away when the door opened up and there were Feynman's crystal blue eyes. He asked, "So what are you doing?" I was deeply involved at that point with primate testing, an enterprise that was intense and costly. Building training devices for each monkey was getting expensive, and the overall data collection and analysis was clumsy. And there was the problem of monkey virus B, a deadly disease that could be communicated to humans by a monkey bite. So when Feynman asked me the question, I had a ready answer.

"Well," I said, "I am trying to build a device that we could implant in each monkey that would send out a radio signal identifying him. Then we could put all the monkeys in one big cage, have a testing platform at the end of the cage, and when the monkey hopped up to play the game, a computer would recognize which animal it was and keep the response data sorted out." Or something like that.

Feynman furrowed his brow and said, "I have a simpler system. Put the monkeys on a differential diet so each weighs a different amount. When they get on the testing platform, let a scale detect which monkey it is by weight and use that as a way to keep track of the data. No fancy radio transmitters and no implant surgeries." He smiled, winked, and got up and left. I was stunned but soon got back to work. A few minutes later, Sperry walked into the lab as he often did. I told him the story, we chitchatted a bit, and he left.

About thirty minutes later Sperry came back. He said, "It won't work." Puzzled, I asked, "What won't work?" "Feynman's idea," he answered. "The animals will cheat and won't weigh in properly. They will grab the cage bar when they swing in to work." As Sperry left, I thought once again that I worked at the greatest place on earth, brimming with smarts and bristling with competition.

WAIT: HOW DOES SENSORY-MOTOR INTEGRATION WORK?

Seeing any one study now, with the knowledge under my belt that I have gained from hundreds of subsequent studies examining dozens of issues, puts me in mind of several brain mechanisms at work during a given test. During those early days, however, none of those mechanisms had been worked out. As Bogen pointed out, the other widely studied patients of the California series did not show the clear results on the block design test that W.J. revealed. What was up with that? What could possibly explain the individual variation of capacity seen in several patients? Individual variation always offers an opportunity to dig deeper into mechanisms, so it was back to work for me.

In the lab at the time, everyone was fascinated with the problem of sensory-motor integration in cats, monkeys, and humans. We knew that each of the disconnected hemispheres best controlled the opposite arm and hand. Colwyn Trevarthen, a postdoctoral student at the time (who had also done his thesis work at Caltech), was carrying out a series of very clever experiments showing that in monkeys, use of one arm/hand responding on a task meant the opposite hemisphere would learn the problem. The ipsilateral (same-side) hemisphere, even though it had equal access to the information, did not learn. Switch the hand that was responding to the task, however, and quickly enough learning took place in the previously oblivious hemisphere. So, each hemisphere is really good at controlling the opposite arm/hand.[24] That made tremendous sense with the underlying anatomy.

The puzzle was, How did a hemisphere control the ipsilateral hand, which some patients seemed able to do? In other words, how did a left hemisphere control the left hand? While W.J.'s individual hemispheres had little problem controlling the contralateral arm and hand, they were remarkably unable to control an ipsilateral arm and hand. This is quite a dramatic situation. Many of the original split-brain stories about two minds being in our skulls instead of one came from that clear linkage between each hemisphere and its contralateral arm. The film emphasized this basic finding, which was immediately evident to anyone testing the patient. As more patients were added to the study pool, however, many began to show good control over the ipsilateral arm as well as the contralateral arm. Yet, even when there was good control of the ipsilateral *arm,* good control over movements of the ipsilateral *hand* seemed to elude the patients. How did all of this work?

It finally became apparent to us why all the patients did not show the same block design phenomenon as W.J., who was uniquely skilled with his left hand. Some of the patients had more control over their ipsilateral hands than did others. It usually took them a little time after surgery to learn how to control the ipsilateral hand, but most of them got really good at it. That meant that a hemisphere, specialized for a kind of processing that could be revealed by, say, arranging blocks, could, after it learned to control the ipsilateral hand, do it with either hand. Thus we could not tell which hemisphere was controlling the movements of the hands. This dual control made it exceedingly tricky to evaluate the special abilities of the left and right brains.

That most of the patients eventually gained control of their ipsilateral arm was evident. It was true for the original series of patients in California and also for the East Coast patients tested later. It took several of us a great deal of research to figure out how. The problem has fascinated everybody. Myers and Sperry studied it in the cat and Trevarthen studied it in the monkey. Everyone took different pieces of the puzzle. I went after a very simple question: How could a monkey, with one hemisphere viewing the world, pick up a grape with the ipsilateral hand? The answer would be revealing.

Again, the overall puzzle was, Why, when it came to carrying out goal-directed behavior, did animals with their brains divided (and sometimes far more extensively disconnected than was ever done in humans) always seem like they were behaving in an integrated way? How, for example, was the left hemisphere of a deeply split rhesus monkey—a hemisphere disconnection that extended deep into the brain, down even into the pons*—able to control its ipsilateral left hand? Part of our problem was that we had made an assumption. Our assumption, to which we were all so committed, was that voluntary motor control originated from a central command center that had to directly connect to particular peripheral muscles. Because of that assumption, what we observed at first made no sense to us. We had been duped by our own false idea, and as we shed the assumptions, the workings of the brain began to look very different from what we had assumed. The notion that there is an "I" or a command central in the brain was an illusion.

I know this is hard to swallow, which is the very reason it took us so long to figure out. There is no one point where the director lives. No homunculus calling the shots. Dozens of studies finally revealed

---

* The pons is part of the brainstem.

the reality: The animals were engaged in self-cueing.[25] There was no command central. One hemisphere was reading the cues set up by the other to produce an integrated, effective behavioral outcome. Suddenly our whole view of how the brain might coordinate its parts was undergoing what amounted to a paradigm shift.

In order to uncover this strategy, we took high-speed films of split-brain monkeys, each with one eye closed, reaching for objects such as grapes. In these animals the optic chiasm was also divided, which means that information presented to one eye went only to the ipsilateral hemisphere. So, if we occluded the right eye (and I did this by various means including a specially designed contact lens), only the left hemisphere could see. Then we filmed how well the two hands retrieved grapes presented to the animal at the end of a wand. With visual information now restricted to the left hemisphere, the right hand, which it controlled, was quick and deliberate in retrieving the much-desired grapes. As the hand moved to grasp the grape, the posture of the hand properly formed in anticipation of retrieving the morsel of food (Figure 12).

When the animal, his vision still limited to the left hemisphere, tried to use the left hand, however, a different strategy was evident. Cueing was active at many levels. First, the monkey would orient its entire body toward the object. The left, seeing hemisphere had control over gross body posture and orientation. It could easily position the entire body in the correct orientation toward the desired point in space occupied by the grapes. As a consequence, through proprioceptive feedback mechanisms, which provide motion and position feedback from tendons and joints, the right hemisphere now knew, in a general sort of way, where the object was located. Then the left arm would reach out in the general direction of the object. The left hemisphere could initiate left arm movements by signaling, using a body movement, to the right hemisphere to "go." As a result, the right hemisphere commanded the left hand to start off in the appropriate direction, which it now knew because of the proprioceptive feedback from the arm. In short, the right hemisphere knew

FIGURE 12. Schematic reconstruction of the slow-motion movie used to examine split-brain function in monkeys. The film helped us determine how a split-brain monkey could control the arm and hand on the same side of the body as the brain hemisphere that enabled it to see the object it wanted to retrieve.

roughly where the object was. Here was the fascinating part: The left hand remained limp rather than getting ready to grasp the object, because its controlling hemisphere had no idea what the object was. The right hemisphere couldn't actually see it, and the left could not control the distal digits of the left hand. As a result, the hand always looked ill-posed for actually grasping the grape—until the grand finale: Eventually the hand bumped into the grape! At that moment, the somatosensory and motor systems of the right hemisphere were cued and clicked in. The left hand snapped to, formed the correct posture, and grasped the grape. It's a lot like when we stick our hand into a dark drawer to pull something out; as soon as we feel it, we know how to grab it.

So, understanding why we did not see the right hemispheric specialization for block design in all the patients is not only comprehensible, but to be expected. W.J. had additional brain damage other than the callosal brain damage that prevented the ipsilateral sensory/motor system with its feedback mechanisms from working well. The other patients were more intact neurologically, and they quickly learned how to work this system. Years later, when I began to study the East Coast series of cases, we saw how immediately after surgery there was poor ipsilateral control of the arm and hand, which later improved after the hemispheres had learned how to cue each other.

It is impossible to keep at a single project 24/7, whereas it is fairly easy to be highly involved with life 24/7. The human split-brain work kept me busy, but not all the time. I had other research projects going, partly because it was possible that the human work would soon wrap up; then what?

While W.J. was the inaugural case for launching the surgically sectioned split-brain era, it later turned out that he was not the most interesting case. After a while we realized that he had quite limited functioning from his right hemisphere. The original studies on W.J., however, did answer basic questions about disconnection effects. We showed that visual information presented to one hemisphere didn't transfer to the other. We explained the more complicated picture of the somatosensory system, and we worked out the capacity and limits of a brain's ability to control the arms and hands. Finally, we were able to show how one hemisphere was superior to the other at carrying out three-dimensional reconstructions, the block design test. And as I say, we did that all rather quickly.[26]

While that work was getting done, we also became interested in exploring the mental capacity of the nonspeaking right hemisphere and finding out what else it could do. Did it have any language at all? Could it solve problems? Could it learn simple games? For more than two years we pushed and pushed on these issues.

It turned out that W.J.'s right hemisphere had only rudimentary cognitive abilities. While his right hemisphere could do what is called a simple match-to-sample test, its success on more complicated tasks was only as frequent as would occur by chance. So when we flashed a picture of a triangle to the right hemisphere, it could point to a picture of a triangle from a set of many possibilities. If we flashed a word of an object like "apple" or anything else, however, the right hemisphere was at sea. The simple capacity to match objects showed that it was functioning independently, but it wasn't functioning at a high level that involved any kind of language. The linguistic potential of the right hemisphere only became evident as more cases came along.

So there we were. We knew that the cognitive capacities of W.J.'s right hemisphere were limited. Before the next cases, N.G. and L.B., came along, everyone got involved with other projects. We were at a point in the overall program where we all remained excited about our discoveries but were asking ourselves how long it would go on. At that time, we had no idea how rich split-brain research would become when more cases were added to the research pool. The new cases would prove to be game changers in an exciting way, but all of that was to come later. From Sperry's perspective, while we had an exciting finding, he wasn't about to relinquish his long-standing commitment to the field of developmental neurobiology. As Bogen pointed out, it was no wonder he was a bit circumspect in the beginning.

Still, what do you do in your spare time? Well, in those days you easily did more research. Money wasn't an issue. Time was there if you made it. And certainly the questions were there—being generated in large part by the successful human work. After studying W.J., one of the first questions I had was, Were memories for events in a monkey's life laid down doubly, one in each hemisphere? Language-based memories in W.J. were clearly present in only the left brain. How would a monkey that had learned various visual discriminations while normal respond to querying each hemisphere for that knowledge after it had undergone split-brain surgery?

And how would I run those experiments? While Sperry's laboratory was a rich community of personalities, it was also a very cooperative community. To get me going, Trevarthen and Sperry did the surgeries. Lois Bird, Sperry's lab technician, helped me learn how to train monkeys. Reggie in the shop taught me how to build testing devices, even though he did most of the work. And my best friend at Caltech, Charles Hamilton, an advanced graduate student, taught me everything else. So they launched me on a stint of primate research of one kind or another for more than fifteen years, all while I continued the human work.

Working with animals, especially monkeys, is a demanding emotional experience. While monkeys can be aggressive and nasty, they usually aren't. We cared deeply for the animals and following a surgery would stay with them until they recovered. We did all of this naturally and without instruction. But in many labs cruelty was built into the demands of the research. There has been a deep cultural shift since then in how animals are thought of, reflected now in lab practices. Today, animal research is run by professionals and is more carefully monitored than the monitoring that goes on in some human clinics.

The results of the experiment asking whether the same memory was stored in both hemispheres were clean and clear. Monkeys that were taught a visual problem appeared to keep only one copy of the memory. Cats, however, appeared to keep two copies, one in each hemisphere.[27] Our testing with W.J. showed that there wasn't any clear indication that the right hemisphere had copies of the memories that were associated with hearing or reading language, memories which the left hemisphere clearly possessed. Yet we knew that both hemispheres could recognize an object, even if naming it was beyond the right hemisphere's abilities. At one level, it appeared that the monkeys and humans were similar, but on another level, humans and cats were similar. Of course, the problem got more complicated as more experiments rolled in.

The puzzle of where memories are stored is alive and well after all these years. In that original study with monkeys, we also showed that the hemisphere that possessed the memories varied. Sometimes it was the left hemisphere and sometimes it was the right. This suggested that there was not an underlying specialization in the brain responsible for the unilateral memory traces seen in humans. Some patients seemed to have bilateral language capacity, while many more did not. Over the years, and as the human split-brain cases added up, the amount of human variation was also notable. How this works is still a mystery.

THE NOBEL PRIZE

It is no secret that Roger Sperry and I had some difficulties later in my career. As the years rolled by, I chose to continue the split-brain approach in my research and published accordingly. Scientists frequently switch their approach and what they study and it aggravated him that I didn't. We had exchanged some unsettling correspondence both in the early 1970s and mid-1980s. While sharing credit was not his strong suit, it also should be no secret that I never had anything but the highest regard for him. When he was awarded the Nobel Prize in 1981 for split-brain work, it was well deserved. *Science* magazine asked me to write an appreciation for him, which I did with enthusiasm (see Appendix I). I much prefer to have this statement be the public record of my thoughts about him than the sometimes rather ghoulish and misleading letters I received indicate.

CHAPTER 3

SEARCHING FOR
THE BRAIN'S MORSE CODE

For every minute you are angry you lose sixty seconds of happiness.
—RALPH WALDO EMERSON

M Y VIEWS CHANGED OVER TIME. THE FIRST TWO REPORTS by Joe Bogen, Roger Sperry, and myself were mainly about W.J. Single cases are always interesting but hardly definitive. As more cases came under study, it was clear that while what we'd learned from W.J. set the stage for split-brain study in humans, it would not define what could be learned. The simple, straightforward picture at the start was energizing but far from complete. In fact, fifty years later we are still far from completely understanding the full neural and psychological consequences of severing the major neural cable in the brain connecting the two half brains. While callosal surgery for epilepsy has always been done infrequently, it has become even less frequent with the advent of other surgical strategies and better pharmacologic interventions.

Joe used to say all the time that science advances first by a major finding being unearthed, then years of other stuff being piled on top of it, smothering it with details that are distracting. From that perspective, the newer cases were complicating. For example, as I mentioned in the last chapter, they quickly revealed that one hemisphere could control both arms but not both hands. We then went on to find that because each hemisphere could control both arms, a touch to either side of the torso could be located and pointed to by either hand. This finding made it seem that sensory information from both sides of the torso projected equally to both hemispheres, which we would then find was not the case. We also observed that the right hemispheres could sometimes seem extremely smart and sometimes even uniquely smart, capable of some nonverbal skills. In short, a much more interactive and dynamic mental system seemed present, even though the interactions were being controlled by two utterly private and disconnected cognitive systems. We slowly began to realize that it was not going to be easy to understand whether we were examining the separate psychological processes of one isolated and disconnected half brain or being duped by the other half.

All of this was happening a couple of years into the Caltech testing program. As still more cases were added for study in Los Angeles (and eventually from different surgical centers throughout the country), the issue of the two hemispheres continuously interacting with one another in complex ways became ever more evident. In the mid-1970s the first new patients from the East Coast broadened our basic understanding even more. That development, however, was almost ten years off.

As the mid-1960s approached, I knew at some point I had to leave the nest. It was an unsettling thought, as Caltech remained my scientific heaven. For a long time, I simply ignored that reality and went about my business. My last couple of years in Pasadena were rich, and I was recently reminded how rich when I came across the series of films I had made. Seeing those films, along with a later video of Case D.R., one of our East Coast patients, triggered fond memories of identifying basic mechanisms in both series of patients.

One of those basic mechanisms had to do with emotions. Emotions color our cognitive states almost moment to moment. More primitive subcortical parts of the brain located below the callosum are heavily involved in the management of emotions, and many of those structures have interhemispheric connections. Could it be that emotions experienced by one hemisphere could be detected by or have influence on the opposite hemisphere?

Those questions began to form as we started to see the differences between W.J. and the second Caltech case, N.G. While studying N.G., we began to suspect that each hemisphere wanted to monitor the other. N.G. could control either arm from one hemisphere. We saw an instance in which one hemisphere, by initiating just the slightest head movement, could cue the other hemisphere with a solution to a task we had requested of the first. In a sense, the two hemispheres were cheating like two kids in a classroom. Once we realized what was going on, the findings all made sense. Imagine being tightly yoked to someone, even though each of you remains a completely independent person, just like highly competent tango dancers. Any slight movement of the head by one gives cues to the partner of exactly what to do and when to do it. Of course, it should work that way. In our studies this meant that with practice the patients were going to become better and better at self-cueing.

BRAIN CUEING IS EVERYWHERE

All of this subtle communication between the surgically separated hemispheres was clearly evident in our cognitive testing. We dubbed it "cross-cueing."[1] Modular or separate systems cueing each other in order to generate purposeful and integrated behavioral outcomes seemed to be everywhere. We detected this early on in both the animal and human split-brain work at Caltech and saw it occur time and again when testing our patients over the next fifty years.

In one of the first observations, I was in the process of seeing if patients who spoke only out of the left hemisphere could name simple colored lights shown to both visual fields. In the early days, we were always concerned with whether or not basic visual information could transfer over from the right hemisphere and be described by the left hemisphere, perhaps via intact subcortical pathways.

During one such study, the patient N.G. displayed our newly discovered self-cueing strategy. The test was as follows: If any colored light came on in the right visual field, which projected to the left speaking brain, then there was no hesitancy; it was quickly named correctly. When a light came on in the left field, however, projected to the right brain, matters changed, though it wasn't obvious right away. If we flashed the color green to the right hemisphere and N.G. spoke the word *green,* and the color had been "green," the patient said nothing else, and we got ready for the next trial. At that point in the test, we did not know if the information about the light had somehow transferred over to the left hemisphere, if the left hemisphere was simply guessing, or if the right hemisphere was actually speaking.

The telltale trials were when the right hemisphere saw a particular color, say "red," and N.G. said the wrong color, say "green." After a few flat-out mistakes, the patient started stating the correct color every time. Somehow she had learned a strategy that made it appear that the left hemisphere could name something only the right hemisphere had seen. She would start to say "gree . . ." but then she would stop and then guess correctly by saying "red." What was happening was the left hemisphere was doing the talking and the disconnected right hemisphere was hearing the "gree . . ." being uttered by the guessing left hemisphere. The right hemisphere somehow stopped the speech emanating from the left brain by giving some kind of cue, such as nodding the head or shrugging the shoulders. The wily left hemisphere would pick up on the cue, which it had figured out during the first erroneous trials, and would change its response to the only other color! All of this happened in the flash of an eye.

I wanted to look into this self-cueing strategy more deeply. In a sense, this kind of cueing was taking place outside the brain. The patients were learning the equivalent of tango dance strategies: That is, one side of the body tapped the other to get communication going between the two halves of the brain. It could look like the two separate brains were unified by internal connections and communication, but in reality, it was external signals providing the communication that united the two. We also began to wonder if there was cueing actually going on within the brain. After all, the surgery only disconnected the cognitive and sensory systems housed in the cortex. There were still dozens of ways one side could connect with the other in complex, albeit more indirect, ways within the intact subcortical pathways of the brain. And, as I mentioned above, we wondered about the more ethereal aspects of mental life, such as emotions. It seemed as though monkey experiments could get at the emotion question.

So, in Caltech style, we just did it. This required building more special testing devices more animals, and more honing of my own surgical skills. Surgical procedures were taken very seriously and carefully planned. We all trained ourselves, first by attending surgeries performed by seasoned members of the lab. I was lucky, for the highly skilled Giovanni Berlucchi, visiting from Pisa, Italy, let me sit in on his surgeries. Sperry was also a fantastic surgeon. Once, when I was in the surgical room, watching him at a delicate point in the surgery, he looked intently into the operating microscope and softly said, "I can't seem to see the anterior commissure." I leaned forward to better hear him and in doing so, jarred the table, to which he calmly said, "Oh, there it is." Always unflappable.

To carry out this experiment, the monkeys, once they'd recovered from surgery, were outfitted with goggles equipped with one red and one blue lens. The colored light filters allowed different visual images to be projected to the separated hemispheres. We wanted to know what would happen to the work pattern of one hemisphere if the other were suddenly exposed to an emotionally powerful stimulus, such as a snake. Would the emotionally provoked half brain dominate or subcortically influence the half brain that was engaged in the simple and emotionally neutral task of visual learning?

The answer was clear. The animals jumped back. The hemisphere that had seen the emotional stimulus, the snake, and experienced the emotion, fear, cued the rest of the animal: Something is wrong! With that gross and unmistakable cue, the animal became agitated and ceased working on the discrimination task and would not return to work, either. Cross-cueing of a kind was evident once again. In this instance it appeared that one separate and distinct mental system could be agitated and in that agitation not let the other mental system function in its normal way. The idea crept into our heads that the "mind" was a collection of mental systems, not just one. At the time, this was a new and important idea. It was absolutely crucial to the understanding of why the split-brain monkeys, as well as the patients, behaved as they did.

Testing different theories on both animals and humans continued. In the later 1960s, years after we both had left Caltech, Steve Hillyard and I were collaborating on a study. We were trying to figure out the language capacity of the third Caltech patient, L.B., when we spotted another variant of cross-cueing. We set up an easy test for the patient. All he had to do was name the number (1–9) that was flashed either to the left or the right visual field. Normally we would expect the right visual field stimuli to be named quickly. Thus, if a "1" or a "4" or a "7" flashed up in random order, the patient's left speaking hemisphere would respond correctly. It did. Each number was named with about the same reaction time.

What initially surprised us, however, was that the right hemisphere seemed to be naming all the numbers, too. What was going on? Was this our first patient to show transfer of information between the hemispheres? Alternatively, was this a right hemisphere that could speak? (This possibility is always there and must always be investigated.) Or was the right hemisphere somehow cueing the left hemisphere again?

Hillyard plotted out the reaction times for each response, and the strategy L.B. was using became apparent. All the numbers flashed to the left hemisphere were named rapidly in about the same amount of time. When the same random list of numbers was presented to the right hemisphere, however, "1" was reacted to more quickly than "2," which was reacted to more quickly than "3," which in turn was reacted to more quickly than "4," and so on all the way up to "9." Another cross-cueing strategy revealed! The left, speaking hemisphere started counting using some somatic cueing systems such as a slight head bob, which the right hemisphere could sense. When the number of bobs hit the number that was presented to the right hemisphere, the right hemisphere sent a somatic stop signal that the left hemisphere could sense. At this point, the left knew that must have been the number that was flashed and said it, not the right![2] Unbelievable. Trying to outfox this cross-cueing system, we ran another series of trials. This time the patient was required to respond immediately. While the left hemisphere continued to respond correctly and quickly, the right hemisphere's score dropped to chance. The brain was shifting strategies to accomplish the same goal.

THE POWER OF THE BEDSIDE EXAM: CASE D.R.

When studying neurologically disrupted patients, certain general principles emerge. For example, the patients nearly always strive to complete a goal that has been set by their examiner. One might think and hope they are solving a task in a particular way, even when in fact they are solving it in another. The challenge is to identify *the way* they are solving it. When you do, underlying mechanisms are revealed that are frequently surprising. When sorting through my hundreds of hours of videotapes of patients, I recently came across a particularly vivid example, which revealed how cueing can occur when a patient is simply trying to copy a gesture made by one hand, with the opposite hand (Video 4).

This was Case D.R., a split-brain patient from the Dartmouth series of cases. D.R. was also a college graduate and an accountant. After spending time in South America, she had moved to New England. Along the way she had become a Trekkie. She had all the *Star Trek* episodes recorded and also owned a rather expensive model of the Enterprise! After her surgery, she showed all the standard disconnection phenomena. Visual information did not transfer between the hemispheres, nor did tactile information. Her left hemisphere was dominant for language and speech; her right hemisphere functioned at a lower cognitive level, being able only to recognize pictures, but not to read. We wanted to examine her motor control capacity. With her eyes wide open, I asked her to hold out her two hands, fists closed; that was the starting position for each subsequent command. I then asked her to make the "hitchhiker" sign with her right hand. She did so instantly. I then asked her to do the same thing with the left hand. She also did that quickly. I then asked her to make the "a-okay" sign with her right hand. Again, she did so quickly. When asked to do it with her left hand, after a slight hesitation, she had no problem.

Here is where learning begins for the examiner when testing neurologic patients. One has to make sure that the task a patient is trying to complete for you is being done the way you imagined it would be done. In this case, I knew, of course, that the patient had undergone split-brain surgery. By the time we worked with her in the 1980s, I knew that there was tremendous variation on how well a disconnected hemisphere could control the ipsilateral hand. Of course, there never was a problem in controlling the contralateral hand, as both the sensory and motor systems needed for such activity were all represented in the same hemisphere. Controlling the ipsilateral hand, however, was a very different story. How did her dominant left hemisphere, which had to interpret my spoken command messages, send them over to the motor systems in her right hemisphere, which controls her left hand? Those motor control systems for the left hand were unquestionably managed by her disconnected right hemisphere. How was information presented to one hemisphere being integrated for use in the opposite, disconnected half brain?

Recall that Case W.J. was remarkably unable to control an ipsilateral arm and hand, while having little problem controlling the contralateral arm and hand from a particular hemisphere. This was quite a dramatic situation. As I said earlier, many of the original split-brain stories about two minds being present in our skulls instead of one came from W.J.'s behavior. Nonetheless, as more patients were added to the study pool, many began to show good control over both the ipsilateral arm and the contralateral arm. Yet, even when patients had good control of the ipsilateral arm, good control over the ipsilateral hand seemed to elude them. Again, how did all of this work?

Back to Case D.R.: In the video, she was making hand gestures with both hands that appeared responsive to my verbal commands. I knew D.R. had split-brain surgery and that her dominant language hemisphere was disconnected from the motor control systems of her right hemisphere. I was eager to learn how she was completing the task of controlling her left hand so easily, given that her two half brains were disconnected from one another. What to do? Armed with this knowledge, I changed the exam ever so slightly, and this allowed the answer to emerge.

Instead of asking D.R. to make a "hitchhiker" gesture with her right hand first, I asked her to make the gesture with her left hand first. She couldn't do it. After she failed, I then asked her to make it with her right hand, which she did instantly. Same story with the "a-okay" sign: She just couldn't do it if the left hand had to do it first. Why was that?

Obviously, what was going on was that when the right hand (controlled by the left hemisphere) went first, it set up a model and an image for the right hemisphere to see and to copy. If a model was present to copy, then the right brain could mimic a gesture and perform that task easily. In essence, the patient had visually cross-cued the information from one hemisphere to the other outside of the brain, thereby trumping the fact that her brain hemisphere connections had been severed. If this were true, then what would happen if the patient were asked to do the task with her eyes closed? With bedside testing this was easily done. The exam continued.

I asked the patient to close her eyes and to make a "hitchhiker" sign with her right hand, which, again, she instantly carried out. Now, with her eyes still closed I asked her to make it with her left hand. Amazingly, she could not do it. The patient's right hemisphere could not understand the spoken command, and with her eyes closed, the left hemisphere could not cue the right hemisphere by providing the model gesture of the right hand to copy. As a consequence, the left hand sat there frozen with inaction.

This one simple bedside test revealed so much. It revealed not only the dramatic disconnection effects of the surgery, but also a basic truth about goal-directed behavior. We are all eager to achieve singular, unitary goals, and we behave as we wish in specific circumstances. We somehow achieve this unitary output from a highly modularized brain with multiple decision centers. In human patients, when the normal neuronal pathways are disrupted, the goal may still be achieved through whatever alternate mechanisms and strategies remain available. In this instance, two things were clear: The right hemisphere, disconnected from the left, could not follow a verbal command, and it was the hemisphere with the major control over the left hand. The explanation might have been, however, that the left hemisphere could have governed the ipsilateral left hand through ipsilateral corticospinal pathways (a small number of neurons that are uncrossed) that we know exist. Yet we had already shown that that explanation could not be true, because the verbal command to gesture with the left hand could not be followed either when the eyes were closed or when the left hand was directed to respond before the right hand. What was going on?

Clearly the right hemisphere could only execute the command when it visually saw and imitated a model of the posture being requested. The overall system, with all of its separate modules, had cued itself into completing the goal. This cueing is a ubiquitous mechanism of goal-directed behavior.

NEW CASES, NEW FINDINGS

While these experiments on basic sensory-motor control were booming along both at Caltech and in subsequent years at Dartmouth, I grew fascinated by the idea that we could show what a separated right, nonspeaking hemisphere could do in terms of thinking, perceiving, understanding, planning, and all the rest. Getting anything out of W.J.'s right brain had proven difficult, even though he was clearly skilled at visuomotor tasks such as the block design test. He responded normally and easily to the flash of a picture or word to his left hemisphere. Yet flashing the exact same information to the right hemisphere generally provoked only minimal response. It was like pulling teeth. Driving to Downey every week in my old Studebaker was becoming a chore. Sometimes I went just to claim a $3.67 gas reimbursement, a rate that enabled me to run my car for the rest of the week.

It wasn't until we began working with N.G., a pleasant young woman with an exceptionally supportive husband, that we moved beyond the fundamental sensory-motor integration tasks we had studied so intensely in W.J. She, like W.J., had been operated on for uncontrollable epilepsy and treated by Bogen and his neurosurgical mentor, Philip Vogel. She was agreeable to testing and, as with most of the patients, it became a big part of her life. After all, here we were showering attention on the patients and compensating them for their time. We all built lasting relationships that have continued over the years. Just last spring a relative of N.G's husband called me after almost forty-five years of no contact, just to say hello.

Soon after N.G. came L.B., a young boy of twelve, another favorite and extensively studied patient. He had also been operated on to control his severe epileptic seizures. L.B. proved to be a remarkable case. Years later, again out of the blue, he sent me a yet-unpublished manuscript he had written about his personal experience as a patient and experimental subject. In writing his personal perspective, L.B. had been assisted by Caltech's science writer, the wonderfully sensitive Graham Berry.[3]

These two new surgical cases brought real energy to the project. While they quickly confirmed the basic disconnection effects of W.J., they provided us with new insights about the function of the right hemisphere. Their right hemispheres responded to our tests with glee and vigor, even though their left hemispheres remained unaware of the content being processed by the disconnected and largely silent right hemisphere.

By this time, I was using my Bolex camera a lot. When testing N.G., I would place it on a tripod pointing down in such a way that one could view not only the patient's face, but also objects placed out of view that the patient would sometimes have to touch as part of the test being run. In this way, it was easy to visualize the actual tests and the sometimes stunning results. First, objects held in the right hand could be easily named, but not those in the left hand. Second, pictures of objects flashed to the left hemisphere could enable the opposite right hand to find the matching object, but not the ipsilateral left hand. Third, and here was where the new era really began, pictures and even words flashed to the supposedly language-impoverished right hemisphere triggered the left hand to retrieve the correct object out of view.[4] It was astonishing and remains so to this day (Video 5). We were looking at the first real evidence that the right hemisphere could be capable of cognitive activity and complex behaviors without the left hemisphere knowing anything about it.

After years of studying N.G. and L.B., Sperry and I concluded that the right hemisphere had a large vocabulary. It could react correctly to printed words as well as to line drawings of all types.[5] There was even some capacity in the right hemisphere for simple spelling and an occasional written word, but rarely so. We pushed on in the hope of finding independent higher-level thought of some kind, and attempted several tests requiring simple arithmetic. Here we had an occasional success with simple addition, but none with subtraction.

We were always on the lookout for brain functions that did transfer from one separated hemisphere to the other. After the monkey work with emotive stimuli, I wanted to see if humans would react in the same way. Would a potentially emotive stimulus cue the opposite hemisphere in some way, where it was clear that nonemotive stimuli did not? This test required a stop at the back of the magazine shop, where, at that time, plain cardboard was placed in front of the risqué magazine covers. Magazines would have to be bought and pictures clipped out, photographed, then put into a slide carousel so they might suddenly appear in the left visual field as part of a sequence of pictures of more routine objects, such as spoons and coffee cups. I was nervous about this experiment. While I was a vigorous postadolescent, to be sure, I was also a practicing Catholic, and as you well know . . .

Somehow I survived the guilt of producing the pictures and ran the test, first on N.G (Video 6). I had the camera all set up. It would catch any facial expression she might have, but since it was back in the silent film days, there would be no voice recording. Luckily the film was clear, and one could see her reacting to my questions.

MSG: Fixate the dot.
N.G.: Okay.

A picture of a spoon was flashed to the left visual field, which revealed its contents solely to the right hemisphere.

MSG: What did you see?

N.G.: Nothing.

Expressionless reaction on her face.

MSG: Okay, fixate the dot.

This time a picture of a nude woman was presented to the right hemisphere.

MSG: What did you see?

N.G.: Nothing.

. . . but she then fights off a grin and finally lets it all out and does a full chuckle.

MSG: Why are you laughing?

N.G.: Oh, I don't know. That is a funny machine you have there.

I was excited with this result, though it would be several years before I fully understood its implications. At the time, I simply wanted very much to confirm it with W.J. A few days later I piled all my testing gear into my Studebaker and drove off to Downey. I flashed several neutral pictures to W.J. before flashing the nude picture to the right hemisphere of this World War II veteran. Again I asked, "What did you see?" With the most expressionless face I think I have ever seen, he replied, "Nothing." I was so disappointed. Maybe the test on N.G. was a fluke.

To be complete, of course, I tested W.J.'s left, talking hemisphere directly. To my surprise, and without expression of any kind, W.J. said "A pinup?" I said, "Yes." As I was fiddling around with my gear, W.J. dryly added, "Is that the kind of co-ed you have at Caltech?" So there you go. Neither hemisphere found the nude engaging. Linus Pauling was right: Never assume anything.

LEAVING THE NEST

I took the phone call from Howard Kendler, the chair of UCSB's Psychology Department, on the hallway phone in the Alles Laboratory at Caltech. Phones were not allowed in individual offices in Sperry's lab, and probably for good reason. When the phone rang, it was sort of an anti-signal: Answering it would invariably interrupt what you were doing. As it was, occasionally, you still had to pick it up and then go fetch somebody. Phones were a pain in the neck.

Nonetheless, it was my turn to take a call. I picked up and Howard said, "We would like to hire you as an assistant professor here at UCSB at the startling salary of nine thousand, five hundred dollars for nine months." A nine-month offer meant you had to find your own salary for the other three months through grants or other means. As in the moment before a car crash, some vital issues flashed through my mind as he spoke. First, like it or not, it was time to leave Caltech. I had been there five years, and new students were coming in to pick up the ball. Second, I had accepted a postdoctoral fellowship in Pisa to work with my dear friend Giovanni Berlucchi. Third, I needed a job when I came back. And fourth, I would be within a hundred miles of Caltech and close to the patients, so I could continue research. Then and there I heard myself saying, "I'll take it!" And that was that.

Of course, the decision was not without pain and a sense of loss. I had grown up into the ways of not only the scientific world but also the social world at Caltech, which included political meanderings. My friendship with Bill Buckley had deepened, and in 1964 he invited me to be his sidekick at the Republican National Convention in San Francisco. The great American author John Dos Passos, who had become conservative after his flaming left-wing youth, was there to write an article on the convention for *National Review.* Bill gave me the assignment of shepherding him. One-eyed, in his seventies, Dos Passos had more energy than six young men. I could barely keep up. It was a sublime experience, capped off by the last night's chore of typing out his piece. It took me all night. The next afternoon, when I saw Bill, who had just been debating Gore Vidal, he calmly said to me, "Mike, I can see typing is not your strong card." Oh well. It wasn't.

I was involved in other political functions at Caltech, including Martin Luther King Jr.'s final visit there in 1965. In the evening, he spoke at the historic Friendship Baptist Church in Old Town Pasadena. It was the first black Baptist church in the city. I was able to get into the back of the church for the sermon, which turned out to be one of the most moving experiences of my life. And there was more. Robert Kennedy came to Caltech in 1964 and writer James Baldwin came in 1963. Meeting such vibrant and driven public figures couldn't help but mature one's thinking about the social world. Baldwin, in particular, was moving. I had the honor of spending an evening in conversation with him in a smoke-filled den at the home of a Pasadena patron. He said that he had moved to Paris years before in order to feel more free as both an African American and a homosexual. When asked why he returned to the States, he simply declared that even though Paris had many advantages, he was an American at the core. (In the small-world department, a couple of years later, Baldwin and Buckley were in a much-reported debate at the Cambridge Union Debating Society in England. Baldwin was declared the winner.)

Now it was time to move on from the lively world of Caltech into my own academic position. Before settling in to that life, it was off to Italy to train with Berlucchi. We had an idea, one so painfully simple and naïve that it cracks us up to even think about it today. The reasoning went like this: Callosum-intact persons can name objects and words in each visual field. Speech is only in the left brain. That means stimuli presented to the right hemisphere by flashing in the left visual field had to somehow transfer over the callosum to the left hemisphere in order to be expressed. By recording in the corpus callosum, we could figure out the brain code! It would be like the Morse code of the brain or something cool like that.

Now, there are lots of reasons to go to Italy, science being only one of them. I have often thought the world should simply turn Italy into something like a national park for the world to enjoy. It is simply stunning, deep, loaded with history, art, and fun; delicious, crazy, breathtaking, irreverent, and hilarious. Forty-five years ago I was about to learn all of this from my first experience in Pisa.

My wife, my two-year-old daughter, Marin, and I were driving down from Paris in a small VW bug and were going to arrive in Pisa very late, around 2 A.M. It was dark, windy, and raining as I sped down the highway. Things were looking pretty bleak and were about to get bleaker: I saw the flashing red lights of the carabinieri in my rearview mirror signaling me to pull over. My heart sank as the officer approached the car. I didn't speak Italian; he didn't speak English. After an exchange of stern pleasantries, he asked for my license and *passaporto*. None of these documents produced a change in the stern expression on his face. He then, as best I could figure out, asked me what we were doing in Italy. I think I figured this out because I heard the word *turista* in there somewhere. Luckily, I had a letter from Giuseppe Moruzzi, the famous Italian neurophysiologist who ran the Istituto di Fisiologia in Pisa. I fished it out and handed it to him. He took it with some disdain and focused his flashlight on it. As he read it, and before my very eyes, he was transformed into a highly respectful servant of the public. "Mi scusa, Professore . . ." I didn't see any reason to correct him on my lowly academic status. Before I knew it, we were on our way with no problems and no citation. I probably could have had an escort if I'd known how to ask. The police respect professors? Wow! I knew then that I loved Italy, my genetic home.

Once settled into a beautiful apartment that had been arranged for us to rent, it was off to work at the Istituto, a beautiful building only a few blocks from our apartment. No room was available in the main building to set up a lab, so Giovanni arranged for us to use a spare building situated in the garden. So there we were. We had some kind of idea and an empty room. It was time to go to work.

Also at the Istituto was Giacomo Rizzolatti, a young neurophysiologist of enormous talent, who later went on to discover mirror neurons (the group of neurons we all possess that track the actions of others). Giacomo and Giovanni (Figure 13) were to become very close friends. Both were superior neurophysiologists, and they were going to try to teach me the trade. It would be a whole new world, a different kind of biology, a time-consuming, demanding, and exacting skill. First, however, we needed to set up for the experiments. In general terms, we needed an operating platform, recording equipment, a projector, a screen, and cats. Berlucchi decided we needed a special screen, a half-dome kind of thing so as the cat gazed ahead, every point on the screen would be equidistant from its eyes. A metal welder in town made stuff, all kinds of stuff, but I can assure you he had never made a half dome out of bent steel for a cat to gaze upon. He seemed game, however, when Berlucchi explained the project, and, although a bit disbelieving, built a half dome. It

FIGURE 13. The Istituto Fisologia in Pisa gave Giacomo Rizzolatti (*top*) and Giovanni Berlucchi (*bottom*) and me space in the garden to carry out our experiments. Both Berlucchi and Rizzolatti became distinguished scientists, recognized throughout Italy and the world.

was delivered on a trailer, drawn behind one of those little scooter trucks you see all over Italy. As it approached, everybody started to laugh. We had a problem. It was too big to get through the door into the lab.

Unperturbed, Giovanni declared, "Non c'è problema." He ordered the dome cut in half in such a way that it could be bolted back together once the pieces were inside the room. It got done. Meanwhile Pasquale, the lab animal technician, found cats for us. Oh, how things were different in those days. Cats did not come from some highly regulated biomedical animal provider, as they have now for at least thirty years. Cats came from the alley! It was Pasquale's job to keep the labs provided with cats. These were no lap cats. These were feral street cats, wild and mean. Even once they were caught and placed into cages, it was a challenge to anesthetize them.

As all of the elements were coming together, Roger Sperry came from Caltech to visit us. He was passing though Italy and stayed with us while in Pisa. As I say, we had the benefit of a wonderful apartment with a guest room, with the only caveat that the guest bathroom's toilet ran a little too long after flushing. Roger, feeling right at home, got the necessary tools, climbed up to the ceiling, and fixed it. He was making sure we were settled in, and we all had a grand time.

Finally, the big day arrived at the lab. Giovanni and Giacomo had perfected the essential operation, which some called the *encéphale isolé* preparation,[6] no slight feat in itself. This allowed the animal to be tested painlessly while awake and staring at the half-dome screen, visualizing the stimuli we projected onto it. In addition, a single electrode could be lowered into the corpus callosum, and we could eavesdrop on the neural signals that were being passed over the callosum between the hemispheres.

Amid great anticipation, Giacomo slowly lowered the electrode into the callosum. As is commonly done in neurophysiology, the recording system was hooked up to a loudspeaker so that the rat-tat-

tat of the neurons firing could be heard. We were ready to hear the Morse code of the brain.

Then it happened. The electrode pierced the callosum. Instead of the rat-tat-tat we expected, the loudspeaker boomed with the excruciatingly clear voice of Ringo Starr singing, "We all live in a yellow submarine, a yellow submarine, a yellow submarine." Giacomo looked up from the cat and calmly said, "Now that is what I call high-order information." Some kind of electronic ground loop had been closed, and we were picking up the local radio station. We all laughed, though we knew this brain code thing was going to be a long haul.

In the end we did complete a nice piece of research.[7] We showed how the individual neurons in the callosum coded for visual information, either to one side of the visual midline or the other. In follow-up studies over the next year, Berlucchi and Rizzolatti were able to show exactly how the callosum enables the actual two halves of our visual world to seem like only one. The Nobel laureate David Hubel describes the experiment and called it the best example of exquisite neural specificity that he knew:

> Having cut the optic chiasm along the midline, they made recordings from area 17, close to the 17–18 border on the right side, and looked for cells that could be driven binocularly. Obviously any binocular cell in the visual cortex on the right side must receive input from the right eye directly (via the geniculate) and from the left eye by way of the left hemisphere and corpus callosum. Each binocular receptive field spanned the vertical midline, with the part to the left responding to the right eye and the part to the right responding to the left eye. . . .
>
> This result showed clearly that one function of the corpus callosum is to connect cells so that their fields can span the midline. It therefore cements together the two halves of the visual world. . . .[8]

ITCHIN' FOR MY OWN LABORATORY

After a few months in Pisa, I was discovering that neurophysiology was not for me. It takes long hours, as all research does, but also great patience, not my strong suit. I was ready to get back and start my own academic life. I missed testing patients, I had a bunch of follow-up experiments I wanted to do, and Santa Barbara seemed a long way from Pisa: I was beginning to feel isolated and out of the loop. I wrote Kendler, the UCSB psychology chairman, and asked him if I could return early and start in January instead of July. He somehow made it work at that end, and my friend Giovanni arranged to terminate my fellowship short by six months in Pisa. Overall, it wasn't my finest hour, but that is what happened.

The excitement of a new job, a new academic rank, a new sense of destiny all colored my early impressions of the University of California, Santa Barbara, in 1967. It is a spectacularly beautiful setting, and the psychology department was full of talented people, including David Premack. Most of the department came out of classic experimental psychology, a whole new world to me. I was so enthusiastic that I asked Sperry if he would consider a job at Santa Barbara. It turns out that this is a fairly common pattern. Sperry actually came up and visited, and I think he even met with the chancellor, but in the end, it didn't work out.

What did work out was my first grant, which I had applied for before taking off for Pisa. I had written it during my last year at Caltech and was able to have it checked by Sperry and others in the lab. It was about both animal and human work that I wanted to continue. Everyone said it was good and wished me good luck. In the 1960s, grants were fairly easy to get, and my good luck continued. Although I hadn't realized it at the time, Sperry was chair of the NIH study section that reviewed the grant. I am sure he had to recuse himself for its evaluation, but it never hurts to have someone on the committee who is knowledgeable about the topic, even if they are standing outside the door. When I arrived at Santa Barbara, I was able to set up my monkey lab quickly and also to start testing patients again.

Most psychologists don't have the luxury of testing patients whose hemispheres are disconnected, a condition that makes the examination of the separated hemispheres relatively easy. Experimental psychologists measure how long things take to do, or how many errors people make when doing some kind of task. From those kinds of observations, they build a model about how such-and-such might work and that, indeed, there is a mental life guiding our behavior. They are really good at it, too. I was surrounded by that kind of expertise at Santa Barbara.

One of the issues the early split-brain work had helped to frame was the question of how information gets integrated into the regularly intact normal brain. When we look out at the world and see a scene, each half of what we see goes to a different hemisphere. Yet, to each of us, it all seems to be one unified scene with no stitching up the middle, no gap in appreciating the left side of the scene versus the right side. How does that work? Maybe all is not gapless. Maybe detectable timing differences do exist and are somehow masked? We made an early contribution to this kind of question using a very simple test.

Undergraduates at Santa Barbara were brought in, and my new graduate student, Robert Filbey, began testing them. Filbey, with his John Lennon glasses and long curly hair, was a wonderful soul and free spirit. His roommate at Pomona College had been Larry Swanson, who went on to be one of the world's leading and most imaginative neuroanatomists. In contrast, after this experiment, Filbey decided that graduate school was not for him and retreated to Garberville, California, to live life as an artist. Over the years, his drawings have graced my books, and his wit is endless. But back then, he was working hard in the lab.

The task was for our volunteers to fixate on a point on a screen that was equipped with both a voice-operated relay and a simple manual electronic key. On the first set of tests, a dot flashed up on the screen after a warning buzzer, and it appeared either to the left or right of fixation. Half of the subjects were told to say "yes" if a dot appeared, and "no" if the screen remained blank, which is to say, if nothing appeared. The other half of the subjects were told the opposite: Say "yes" if the screen remained blank (nothing appeared) and "no" if a dot appeared. The results were intriguing.

When the dot came on in the left visual field or when it was a blank trial, the verbal response, which had to come from the left hemisphere, was 30 milliseconds slower than when the dot was presented in the right visual field. Thus it appeared that when the dot was initially projected to the left, speaking hemisphere (from the right visual field), the overall response was much faster. But when there was a blank trial in the left visual field, the response was slower, because after all, the fact that nothing was presented had to be deduced by the left, speaking hemisphere, and[9] the left hemisphere must have been waiting for the right hemisphere to report in, and that took time.

Still, this lengthy response time (30 milliseconds) didn't make any sense. We knew from physiological studies that it should take only 0.5 milliseconds for a nerve to conduct a message across the callosum. Why were our behavioral measures so slow, and if they were that slow, why didn't we sense it in some way in our own experience? We did one more run at it. Instead of requesting our undergraduates to say "yes" or "no" to the flashed dots and blanks, we instructed them to respond with their right hand on a lever that was to be pushed one way for a "yes" response and the other way for a "no" response. Changing the experiment in this way might reveal that we were not dealing with a transfer time from one hemisphere to the other, but with the simple fact that the right hemisphere was slower than the left when asked to respond to even simple tasks.

Well, the undergraduates gave a clear answer to the experiment, which continues to puzzle researchers to this day. The manual response to a "dot" was equally fast when the right hand was responding, no matter in which visual field it appeared. This suggested that each hemisphere could organize a motor response equally fast. What was surprising, however, was that it took 40 milliseconds longer for the hand to respond to a "blank" trial. Wherever in the brain the decision to respond was being made, it took longer for both hemispheres to report that a "blank" trial had occurred. Perhaps it took longer for a brain to decide that nothing had happened than that something had. Maybe we were not getting at the crux of the matter after all. That happens in science. It happens most of the time.

Thankfully, others have worked on this problem and made real progress. In fact, an entire Italian research team worked on it, led by Berlucchi.[10] Indeed, we continue to work on the problem of how the two half brains coordinate their activity and functions. While it seems on the surface to be a split-brain specific issue, it is really dealing with one of the central issues in brain research: How do parts of the brain interact with other parts? As the late comedian Henny Youngman would say, "Timing is everything." In the brain, usually the parts are only a few microns away from each other or a few centimeters. In both cases, local processes have to conduct their business and then either send or somehow coordinate the information with another part of the brain. Asking how this works between the hemispheres gives the researcher a bit of breathing room, as the physical distances between processing sites are very far apart.

DAVID PREMACK AND
HIS FORTY-YEAR-OLD QUESTION

In Santa Barbara many new lines of research were born and more lifelong friendships were established. One of my closest lifelong friendships has been with David Premack, and it began there (Figure 14). It is difficult to think of a living psychologist more influential than he has been. When we consider our origins, our history, our uniqueness as humans, it is Premack who has been our best guide in the understanding of who we are. As Liz Spelke, the distinguished Harvard psychologist, once told me, "Oh, Dave discovered everything first."

Before his pioneering work on the cognitive and the "possible" language capacities of the chimpanzee, Dave made fundamental insights into the nature of motivation.[11] Behaviorism had developed the view that animals were motivated by external contingencies, a view that failed to consider that animals might have internal states and preferences. He turned the entire view about the nature of reinforcement on its ear by looking beyond what was easily observable. Using the methods of science, he unearthed the underlying principles of what motivates living creatures to act.

He tested these principles by challenging the minds of chimps and, in particular, a chimp named Sarah. I know, since she lived down the hall from my office for years. I don't care for chimpanzees. I have always found them too aggressive and bestial and, quite frankly, would walk in the other direction when Sarah approached with her trainer, Mary, or with David.

Even so, Sarah was no ordinary chimp. She was exceptionally smart and engaging. She was also volatile. Dave perfectly managed her by being even more unpredictable and clever than she had ever seen a human be. This was one *Homo sapiens* who always beat her at her own game. Dave had established a social relationship with her, and then began to use it to explore exactly what was, and was not, in and on Sarah's mind. At that time, Dave

FIGURE 14. David Premack, a true creative genius, with one of his friends. Through his research on animals, Premack changed how we think about the mind.

was beginning to clarify the intellectual limits of our closest living relative, and in doing so, he began to unearth the factors that make humans unique.

Premack's work did not go unnoticed by the University of Pennsylvania. Before I knew it, David and his wife, Ann, were on a plane with Sarah, headed to Honey Brook, Pennsylvania. A chimp facility was being specially built in Amish country to house the chimps for his research. It was there, with the help of Sarah and a small group of young chimps, that Dave gave birth to the idea of "theory of mind," one of the major ideas of twentieth-century psychology. TOM, as it is called, reflects the ability of one's mind to attribute mental states, such as beliefs and desires, not only to oneself (I believe cats are devious) but to others (he wants to get a dog). So we may have a theory about what a chimp and our dog believe and desire (I think that Fido wants to play ball), but do a chimp and a dog have a theory about our beliefs and desires? (Does Fido think for a moment that I am sick of throwing the ball?) Does the chimp have a theory about other chimps? Does it understand that others have thoughts, beliefs, and desires and some primitive understanding of what they themselves are? As in all breakthroughs, it is the ingenuity of the question that makes the impact. This is another Premack specialty. Dave has the rare ability to "turn an issue on its ear." The question of whether or not an animal could have a theory about humans (or anything else) did exactly that. He changed our perspective and opened up a wealth of research on animals, children, and various neurologic syndromes. While an argument rages on in the animal research literature about the capacity of chimps, or orangutans or dogs for that matter, it is clear they don't have much theory of mind, and certainly not to the degree that humans do. It should come as no surprise, therefore, that this question would have a huge effect on me and how I was to think about issues in the next phase of my own work.

PICNICKIN' WITH STEVE ALLEN

One dimension of working for a university that is missing from a medical school or a place like Caltech is that one has the duty to teach undergraduates. Some people thrive on these assignments and are truly brilliant at the task; some are not drawn to it and are uninspiring. Still others find it takes too much time away from their research. I fit into the latter category. It is true that by teaching, one comes to understand a topic more deeply, especially topics about which one has only a passing interest, but when one is committed to a particular area of research, most topics are of that kind.

Once I was established as an assistant professor, I had Psychology I duty (introductory psychology). Almost one thousand students showed up three times a week for class at the university's largest auditorium, Campbell Hall. A bit of crowd control and mass motivation are required when the numbers are this big. In short, one has to keep the students interested and motivated and entertained. Doing it three times a week was grueling. Like everyone I know who teaches these large courses, one seeks some kind of relief—movies, guest lecturers, and more. I had an idea to have a guest lecturer to beat all guest lecturers, my new best friend, Steve Allen. So I called Steve and asked if he would like to come up and give a lecture . . . gratis. He accepted in a flash. The lecture was at 8 A.M., so he drove up the night before, stayed at a motel near the University, and showed up bright and early.

We had agreed he should talk about the creative process. Steve was a man of many talents, songwriting being one of them. Sitting at the piano I had had moved into the lecture hall, he told the story of what was behind his greatest hit, "Picnic," the theme song from the movie of the same name. Steve had written the lyrics in record time after receiving a call from the producer. As he looked back on it, he basically came up with what a psychologist would call a "resource allocation" model for creativity. Usually, he said, he would be asked to write a song with absolutely no constraints put on the assignment. For the song "Picnic," however, the producer said, "I want you to write the lyrics to the theme for our movie, which stars William Holden and Kim Novak, and they will be dancing at a picnic." As Steve pointed out, all his energies became focused on the task at hand, and it was quickly accomplished. In contrast, in the unconstrained situation, so much energy is lost trying to define the context and idea for a song, one is depleted by the time the actual task is under way. It is much more effortful and takes longer.

While the visit was a terrific success, it underscores how complex life as a teacher and researcher truly is on a day-to-day basis. Preparing for a lecture, if it is to be any good, takes time. Doing good research takes time. The two enterprises invariably clash. Sperry used to complain about it and always told the story about what Karl Lashley, one of his famous mentors, had told him during his postdoctoral years at Harvard. "Don't teach. But if you have to teach, teach neuroanatomy. It never changes."

Of course, there are some colleagues who possess incredible verbal memories and have an easy verbal fluency that finds them talking all the time. Teaching is no big deal for such people because the entire day is one big verbal flow, interrupted by demarcations of time called "class." It is easy for them. For others, it is a challenge. To this day, I find teaching grueling. A few years ago, when I was asked to prepare six consecutive lectures for the Gifford Lectures Series at the University of Edinburgh, I worked for two years to get them ready. How did I ever do three a week?

SHARING RESOURCES: THE HEART OF SCIENCE

On the scientific side of the Santa Barbara ledger, I received a call one day from a graduate student at Berkeley who wanted to test a split-brain patient on a visual task. Little did I realize it would turn into another lifelong friendship and instigate a new tradition in the lab. The student was Colin Blakemore, now professor of neuroscience at Oxford, as well as at University College London. He and another student had an idea about seeing depth at the visual midline. I invited him down to Santa Barbara, where he set up his gear and tested N.G. This is how it goes in science. A story gets going on how things work. Slowly, with lots of people contributing, it gets added to bit by bit. In this case, Mitchell and Blakemore showed how important cortical connections were to the integration of information isolated in each hemisphere that was responsible for seeing depth at the center of the human gaze.[12] Our early reports on W.J. were being expanded upon, while also illuminating an observation of crucial interest to the specialized field of visual science. The importance of pathways, of discrete neural fiber systems for transferring information around the human brain, had begun to develop.

This very positive experience inaugurated what became a long tradition of bringing outside scientists to our labs. Split-brain patients were a precious resource for understanding brain mechanisms. While surgical medicine was helping people's lives, we were helping people in a different way by trying to understand how the human brain worked. Clearly, we didn't have all the good ideas. The Mitchell and Blakemore studies became classic. It is the duty of scientists to facilitate good science.

I continued to work out more details of how one hemisphere could control the ipsilateral hand.[13] Moving from ideas generated by the human work, I could test specific mechanisms in the monkey and vice versa. By the time I had left Caltech, it had been fairly well established that cross-cueing was a powerful mechanism. Tests on split-brain monkeys and human patients made it clear that cross-cueing strategies could overcome elimination of the major neural communication pathway between the two hemispheres. It is the neurobiologic equivalent of people of whom it is said (and it is said of me): "If you throw him out the front door, he will come back in through the window." During the Santa Barbara years, several more studies on monkeys and the patients confirmed this important idea. Block any way the brain communicates with its modules, and it will most likely find another strategy to achieve the goal at hand.

Life was good. Then, suddenly, I was asked by the senior leadership in the department if I would serve as chair. "What?" I exclaimed. Yes, they all said, we have promoted you to tenure and associate professor after only two years on the faculty. They told me Sperry and other Caltech faculty, such as James Bonner, had written me grand letters, and UCSB had approved the departmental request. I was enthusiastically encouraged to take on the task by all the elements of the department and, in fact, was wooed into accepting the job.

Now that I was tenured, I had to do my share of administrative work. Robert Zajonc, the late distinguished social psychologist at the University of Michigan, once told me that he had successfully avoided administration for thirty years. Finally, out of some sense of gratitude, he took over as director of the Institute for Social Research. "Mike," he said, "it is amazing how right in front of your eyes, friends you have had for thirty years become assholes." He nailed it, and it certainly was true at Santa Barbara. Senior friends of mine would come into the chair's office, close the door, and start twisting my arm for this or that. In many ways it was hilarious, and is made even more so with the distance of time.

Of course, on the fun side, when one takes those positions, little pots of money can be found and applied to good academic causes. I found some funds and arranged to hire Donald M. MacKay, the distinguished neurophysiologist/physicist and student of the mind/brain problem. His home base was the University of Keele, in the United Kingdom, and he was extremely popular with the American neuroscience community. He had written widely on a model of the brain that suggested that even though the brain was as mechanical as clockwork, as he put it, there was still free will.[14] He and Sperry had a running but cordial argument about the issue.

At any rate, I brought him out to UCSB for a few months. It is hard for me to lavish enough personal praise on Donald and his wife, Valerie, as well. We rented a house for him and his family on a sunny hillside that was equipped with a glistening swimming pool. Our families would meet there for barbecues, and one day in particular, Valerie noticed my oldest daughter, Marin, was at the bottom of the pool! Thankfully, Valerie saw her and jumped in and basically saved her life. Accidents happen in a flash, and since then, I have become vigilant whenever kids are around.

When one goes to the extra effort to create a more interesting intellectual community than the status quo, one wonders if it is worth it. Convincing people to think about matters outside their specific interests is a battle and always has been. Fortunately, people who enjoy doing this tend to find each other. To my enormous surprise, the MacKay lecture series brought up to UCSB a young philosopher from the University of California, Irvine, Daniel Dennett, who has become one of the world's great intellectuals. Dan's lifelong interest in free will surely was encouraged by those talks. He drove up for the lectures and, as it turned out, we too have become lifelong friends. All of these seeds count, and they all add up. Fifty years later, I gave a talk at the Vatican on the subject of free will, and it featured the work of MacKay, Dennett, and Sperry.[15]

Feeling my oats and finding other funds to spend, I brought in other famous neuroscientists, such as Brenda Milner, from Montreal (who was the first to study H.M., the most famous memory patient in cognitive neuroscience), and Hans-Lukas Teuber, the charismatic head of brain research at MIT and the founder of its brain and cognitive sciences department.

Linda and I had moved from the tract house that Cliff Morgan helped us buy to a new redwood house on Mission Creek. It was unfinished but gorgeous when we bought it, and we finished it ourselves, calling upon family and friends for help in design and construction. It turned out to be a magnificent tree house sort of thing, with vaulted ceilings, surrounded by fresh redwood paneling, brick, rock, and glass. It was the perfect house for parties, and parties we gave. One of the first was for Teuber. Being the natural pedagogue that he was, he took me into my bedroom, sat me down on our bed, whipped out a manuscript I had submitted to the journal he had helped start, and gave me a lesson in editing! Hearing the clinking of glasses out in the living room along with much laughter, I remember thinking to myself, I thought I worked hard. I am punk.

In June 1968, Robert Kennedy was assassinated in Los Angeles. I was taking a shower in the unfinished part of our house and my wife ran in very upset and agitated with the news. Her command was "You have to do something!" Having met RFK a few years earlier at Caltech, I found this news especially wrenching. I quickly realized I could do something, and sure enough those extra funds were helpful once again. I put together a quick meeting on the nature of violence, calling upon some new friends and some old friends. I called Leon Festinger, my budding new friend. He called his best friend, Stanley Schachter at Columbia, and Ken Colby at Stanford. They called their friend Paul Meehl, at the University of Minnesota, who agreed to come because their friend and former student, Dave Premack, my colleague, would be there, too. I also called Bob Sinshemier, who was chair of biology at Caltech.

While there was some ribbing about Leon engineering a meeting at my expense to see his old friends, the event itself probably hooked me on the value of getting smart people in one space and letting them chew on almost any topic. I didn't realize it at the time, but I was putting together my first interdisciplinary forum. Seeing a truly distinguished molecular biologist, Sinsheimer, and the whiz computer scientist Colby, who was also a psychiatrist, discuss societal issues of violence with social psychologists changed the game for me. Not only did it reveal a new level of conversation that I loved and have sought ever since—and from which I learned that it is only the really smart people who let everything be put on the table—but it also revealed how productive interdisciplinary discussions can be.

My interest in public affairs continued to grow. In the spring of 1969, I wrote an op-ed for the *Los Angeles Times* on the problem of crime prevention.[16] The Santa Barbara meeting on violence had pulled together some telling points that seemed to be worth repeating to the public. Stan Schachter had reminded the group that roughly 60 percent of felons returned to crime after serving in prison. Nothing seemed to change that number. David Premack had reminded us about the nature of reinforcement and punishment: These were not discrete categories, they were on a continuum. One man's punishment may not be another's. Did the 60 percent who returned to prison not find the experience so aversive? The trick would be to find a punishment that worked for each individual person. When you think about this scheme, it is value free. Look at a group of one thousand people. Some start to engage in antisocial behavior. The goal to constrain them should not have at its core the idea of retribution and justice. The goal should be to pick a punishment that reduces the frequency of the antisocial behavior. It's a big idea that, to this day, continues to be battled over in the legal and scientific community.

The op-ed piece caught the eye of none other than the former governor of California, Pat Brown. Soon I was dining with him at the Beverly Hills Hotel, which was next to his Los Angeles law offices. He invited me down to discuss the whole idea and by the end of the lunch had asked me to take over a floundering book project he had launched with his ex–press agent. The book was all about law and order and Pat's historic role in those issues as district attorney, attorney general, and then governor of California. As I took on Pat's book project, the winds of change once again began blowing my way.*

---

* I did finish the book to Pat's satisfaction and to the publisher's but not to that of Jerry Brown, his son, then just starting his climb to the governorship he'd win six years later. The odd thing about this project is that while it was concluded forty-three years ago, only six months ago a total stranger wrote to me to ask if an old manuscript that he had found was in fact the very same book. Indeed it was. As I reread it, it was chilling to see how little my view—then placed in the governor's first-person narrative— has changed even after recently working for four years on a $10 million MacArthur Foundation project dealing with the same issues.

PART 2

HEMISPHERES TOGETHER AND APART

CHAPTER 4

UNMASKING
MORE MODULES

What is a friend? A single soul dwelling in two bodies.

—ARISTOTLE

I T WAS STILL EARLY DAYS IN THE WORLD OF SPLIT-BRAIN research. The basic findings were now well reported, and the early buzz was that, following a relatively simple neurosurgical procedure, two minds, each with its own set of controls, could exist in one brain. The dramatic disconnection effects had held up for years. Activities isolated to the right hemisphere were unquestionably independent and outside the realm of awareness of the left hemisphere. Clear, clean, and simple. It was the talk of cocktail parties around the world.

Looking back with the benefit of time, the curse of present knowledge rears its head. We all now know about left brain/right brain thinking. We are almost bored with it, just as we become bored with often played tunes. What gets lost is that, at the time, it was a huge deal, in large part because of the then current revolution in psychology. The ideological backbone of American psychology, behaviorism, was dying a slow death, and intellectual centers across the country from Harvard to Caltech were waking up to the fact that cognition and the mind itself could be studied. Behind much of this thinking was Lashley and his stance that mental properties could be studied by examining the neurophysiological processes of the brain.[1] He adopted the term *neuropsychology,* which in his day meant the brain processes of the normal brain rather than the dysfunction of the brain due to lesions or injury. Ironically, he liked Akelaitis's finding that cutting the callosum seemingly did nothing to disrupt the brain, because, in Lashley's view, it was the "whole" brain that created the mind, not particular parts. Even though he helped launch the modern fields of psychobiology and neuroscience, which are dominant fields today, he would have been stunned with the basic "split-brain" findings of two minds in one head.

There was no getting around it; the new work on split-brain humans was haunting. Our most precious sense of life is our very own private subjective experience—that feeling of *my* mind, which is what we all mean when we think about minds. We all think we each have one, and I mean *one.* To suddenly think it can be divided, that *two* minds are coexisting in one cranium, is almost not comprehensible. To think W.J. had two minds gazing out at the world, two listening and two thinking about others—indeed, two thinking about me—was unsettling. The uneasiness we feel about the idea of having two or even multiple subjective states may well have been what led us to our discovery, years later, of the "interpreter," the special device in our left brain that gives our actions one narrative and the sense that we have but one mind.

It was also clear that the right hemisphere and the left hemisphere did different things. The left hemisphere was chock full of speech and language processes. While the right seemed mute and language impoverished, it was able to do some fancy visual tasks. Those findings gave birth to "mind left" and "mind right" thinking, and suddenly everybody was a neurologist at those cocktail parties.

And, again, it was simple: The left did this and the right did that. The brain seemed simple and organized in big, functional units that were managed by specific regions of itself. The idea took off like wildfire.

By 1969, it was also clear that the two half brains could develop clever ways to interact, making it look like the hemispheres were not disconnected at all. The brain was like an old couple who had lived together for years and finally worked out a way to live together yet be separate. This made it difficult to do the research. We wanted to understand how the brain was truly organized, not just how it had figured out behavioral strategies for seeming connected and integrated. At the same time it was becoming apparent to me that it was in those strategies that we would learn something basic about how the brain was organized. It was the brain's catch-22. We had to be as wily as the brains we were studying. It forced us to be constantly vigilant for its tricks and to continually think up new ways to investigate the patients.

It was during this time that the simple descriptions of brain function were beginning to lose their appeal to me. If the right brain was a separate mental system with at least some language, why didn't people who suffered aphasia following left-brain damage recover more readily and easily? In short, why didn't mind right cover for mind left, just like the left or right kidney covers for the other if damaged? I knew that if I was going to make any headway on this question, I needed to start being associated with a medical center that saw a wide range of neurological patients.

Stepping onto the next lily pad in life is always hard, especially when the current one is holding up just fine. The tug-of-war between taking risks to do something new and staying with the tried-and-true seemed ever present. While I think all of us usually prime ourselves for new possibilities, it is others who bring them to us. In the midst of all this work and rumination, I had been offered a job at New York University.

EAST COAST FEVER

One sunny spring morning in Santa Barbara, I was sitting with Leon Festinger on my redwood deck, surrounded by oak trees and boulders, soaking up the beauty of Mission Canyon. . . . Leon, who had just decided to move to New York City, said to me, "You know, people feel differently about living in New York. Some think it is like Paris, and others think it is like hell." Unable to contain myself while he went on about New York, I said, "But Leon, what do you think of this *house*?" He looked around, noting all the fine woodworking, the lofty ceilings, the stone fireplace, the magnificent setting, and said, "Well, if you wanted this sort of thing in Manhattan, it would cost millions of dollars." Somehow, between his wit and his flattering interest in me, he managed to set the hook. I thought to myself, He takes risks. He is moving from *Stanford* to the New School for Social Research. So, why can't I?

It was time to go east to escape the clutches of Southern California. Leon was the draw, New York was the unknown, and what the future would bring was anybody's guess. Sperry had made it increasingly clear through various surrogates that he didn't want me testing the Caltech patients anymore. While unhappy about that, I understood in a way. He was kicking me out of the extended nest and letting others have their day. In the end I thought that was reasonable, and away we went: We sold my redwood palace, packed up, and moved to New York.

Festinger was the intellectually intense discoverer of "cognitive dissonance," the idea that when a personal belief is challenged by new information, we tend to ignore the new information in order to reduce mental conflict. Leon and I had taken an instant liking to one another a year earlier at his home in Palo Alto, where he held his Stanford graduate seminars. He'd called out of the blue and wanted to know about my research. That meant I would give a talk to him and his class, which he always held in his living room. Leon stuck me in a chair at the front of his living room, sat right next to me, chain-smoking his ever-present Camels, and started to query me on each point of my research. When Leon moved into a new field, he wanted to know everything in detail. He was a fearless explorer of new intellectual territory. In the years that followed, he switched again to the field of archaeology and prehistory. At the time of his death, he was working on the impact of the introduction of technology on medieval societies. I was right behind him on all of these topics, always amazed at his erudition and energy. It was like having your own personal scholar opening for you vast horizons for thought and analysis.

It would be difficult to imagine two more different people—we had different philosophies, different styles, different aspirations, and were of different generations. Our enduring friendship developed around a mutual love for good ideas, good food and drink, and lively conversation. Once we both moved east, we were surrounded by the magic of New York, which included the élan of scientists like Festinger and my political buddy, Buckley. With these intellectual giants, everything was always on the table. Parochialism and simplistic zealotry were to be checked at the door. When it wasn't, my eyes learned to glaze over. As Walt Whitman observed:

> Lives of great men all remind us
> We can make our lives sublime. . . .

My move to New York was part pull and part push. The pull was easy to identify. The New School for Social Research was an unlikely place for Leon to go, given his new intellectual move from the field of social psychology to that of visual perception. He quickly formed a "front" group, fondly called the Inter-University Consortium on Perception, an administrative device to have his friends from Columbia, New York University, and the City College of New York meet to discuss issues in the area of visual perception, and of course to also meet and have drinks. NYU recruited me in the eventful spring of 1968. While in the office of a future colleague at NYU, I remember hearing on the radio that Martin Luther King had just been assassinated. Vietnam was a big issue, and New York was buzzing. Before I knew it, I was part of that buzz. I had moved my family to the Silver Towers at NYU. Out of our twenty-sixth-floor apartment we watched the World Trade Center being built.

My resistance to life in New York was less tangible. For almost eight years, I had lived with an identity—that of being completely immersed in human split brain research. I had been heavily involved for the first five years in virtually all of the experiments: all day, every day. I had written up the work in review articles and was also putting the finishing touches on my first book, *The Bisected Brain,*[2] which Arnold Towe, from the University of Washington, had invited me to write. Still, relinquishing what I knew to be a vibrant way to approach understanding the brain was tough. While the cues were loud and clear, the path forward was not. I mean really, New York City?

Being raised in Glendale, California, a place Evelyn Waugh called "home of Forest Lawn, the Disneyland of Death," followed by four years in the boonies of Hanover, New Hampshire, followed by five years in Pasadena and finally three years in Santa Barbara, did not necessarily prepare one for New York City. When we arrived from California, it was a boiling hot, 94-degree August day, complete with humidity, traffic jams, and the local tourist attraction: rudeness. I muttered to myself, "Welcome to New York, sucker. You really blew it." And it only seemed to get worse.

We found a small Catholic school, St. Joseph's Academy, for my daughter Marin, in Washington Square. I can remember picking our way around street bums as we walked to school every morning. One goes from indifference to social surroundings to vigilance. The saving grace of it all was that my daughter didn't seem the least bit concerned. Nor was her younger sister, Anne, back at the play yard at the Towers, where various reprobates regularly came to relieve themselves. The New York mothers simply moved their bassinets and looked the other way. So long as my daughters shared punching the elevator buttons at the end of the day, all seemed well. We grew to love the city, and I spent the next seventeen years living either in Manhattan or in the suburbs. Joan Didion once wrote, though, that no Californian ever unpacks their bags when moving to New York, even though many wind up staying there for decades.[3] It certainly played out that way for me.

THE NEW YORK LUNCH

Any kind of sequence in life appears less like a linear narrative and more like what happens when you beat up batter for Yorkshire pudding. In the pudding there are lots of little bubbles that combine to make bigger bubbles in the sticky goo and grow until the bubble pops, when, of course, it starts all over again. One might work away at something, and then something unrelated might interrupt that work. Or someone enters your life with a whole new set of ideas that set you on a different course. Our brains yearn for interruptions, even though we are annoyed when someone interrupts us.

As we all surely know, life is not a sustained upward climb, where everything and everybody only gets better. After hearing multiple older professors say they were *now* doing their best work ever, a statistician friend asked me, "How could that be?" He found it funny that people actually think that way. The fact is, life's successes and failures are sporadic, and their causes are difficult to determine. Hard work and luck are behind most successes, though it is hard to say, for any given success, how much of each there has been.

Still, we develop an overarching narrative about ourselves and about the issues we study. The narrative keeps us bounded. It allows us to avoid becoming dilettantes. We learn to know and loathe bullshit when we hear it. Being part of the "one brain, two minds" story has been life defining for me. But in 1969, and for the prior three years at UCSB, I had been listening to talk about motivation and reinforcement and dozens of other psychological concepts that had not been discussed much at Caltech. Apparently they had seeped in, and as I hit the deck at NYU, I began research on brain mechanisms underlying reinforcement on rats as well as on split-brain monkeys. No one paid much attention to the work, but I thought it was pretty cool. I was broadening my intellectual interests and no longer singing only one note about split-brain research. Starting new work on neurological patients would have to wait.

So, while I was launching the new scientific work at NYU, the real continuity in my life was provided by Leon and our weekly lunch, which became a twenty-year tradition (Figure 15). We usually ate at a small Italian restaurant on Twelfth Street and University Place, Il Bambino. You could get a couple of martinis and shrimp scampi for about ten bucks. Or we ate at Dardanelles, the landmark Armenian restaurant just down University Place before Eleventh Street. The drill was always the same—a couple of drinks, a great meal, and always, always, lively conversation with one of the smartest men in the world. While noonday martinis are out of favor, once a year my wife, Charlotte (whom I married after Linda and I parted a few years after moving east), and I go out and give it a whirl for old times' sake. Needless to say, we are out of practice and nap immediately afterward. Unbeknownst to me at

FIGURE 15. Leon Festinger and our haunt on Twelfth Street in New York City. The restaurant shown here replaced Il Bambino, our favorite place. Our mutual friend Stanley Schachter, who frequently joined us, and I put together a book of Festinger's work after his death.

the time, Leon used to go back and also take a nap, while I went back to work chewing mints. Leon taught me what friendship means. I have written elsewhere about what it was like to know him and will only reiterate here a small part of it.[4]

Once consumed with the desire to understand an issue, Leon couldn't divert his attention from it, and somehow nothing else seemed important in comparison. It wasn't that he wasn't aware of his surroundings—after all, New York was his Paris—but none of that mattered when he pursued an idea. I know this because early in his career he had left New York for Iowa City, Iowa, to study with Kurt Lewin, and New Yorkers don't move to Iowa lightly.

Lewin was a commanding figure in psychology, and to hear Leon tell it, was adept at generating new frameworks for studying psychological mechanisms. Leon had read Lewin as an undergraduate and was drawn to his ideas. The great philosopher R. G. Collingwood noted in his autobiography that as a very young man, he had stumbled upon the work of Kant. Though he couldn't quite say why, Collingwood sensed that the work was important.[5] For the undergraduate Leon, it was Lewin's work that was so intriguing. He was fascinated by the idea that events could be better remembered if interrupted in their execution. Lewin's research, prior to Leon's arrival in Iowa, had laid the groundwork for the ultimate rejection of the classic laws of associationism, the belief that mental life resulted from the simple associations of events and experiences—an idea with tremendous surface validity.[6]

By the time Leon moved to Iowa, Lewin's interests had begun to shift toward social psychology. So did Leon's during their years of collaboration, even though neither of them had ever received formal training in that area. You want to learn something? Go learn it. The bright, creative mind doesn't need a training program. Announcing that he was now a social psychologist, Lewin took a position at MIT and started the Center for Group Dynamics, which Leon joined. He became interested in the behavior of small groups. Most important, his new group at MIT had developed ways to study complex human decision making in the laboratory. Lewin, Festinger, and many others migrated out of the dust bowl of empiricism to the east to test their hypothesis that private mental states were influenced by group dynamics—the special behaviors that arise both within groups and between groups.

Leon's most famous work started with a small grant from the Ford Foundation to study and integrate the work in mass media and interpersonal communication. He and his colleagues took on the project, and to hear him tell it, the seminal observation came from considering a 1934 report about an Indian earthquake. The fact that puzzled them was that after the earthquake, the vast majority of the rumors that were circulated predicted that an even worse one was coming. Why, after such a horrendous event, would people want to provoke further anxiety? Leon and his colleagues concluded that it was a coping mechanism that the Indian people had developed to deal with their present anxiety. In other words, since the earthquake had filled the population with grief, they had formulated an even greater future tragedy, which cast the present moment in a better light. It was out of this basic observation that the theory of cognitive dissonance was born. It would take seven years of hard work to nail down all of the parameters of the phenomenon, but nail them down he did.

Leon carried out one of his early experiments with his two close friends, Stanley Schachter and Henry Riecken. Although a real group of people had been studied, he chronicled the story with a fictitious character and place. As the story goes, the group had come to believe the flood prophecy of one Mrs. Marian Keech. Months before the day of the flood, the following headline and news report had appeared in the *Lake City Herald:*

PROPHECY FROM PLANET CLARION. CALL TO
CITY: FLEE THAT FLOOD. IT'LL SWAMP US ON
DEC. 21, OUTER SPACE TELLS SUBURBANITE

> Lake City will be destroyed by a flood from the
> Great Lake just before dawn, Dec. 21, according
> to a suburban housewife. Mrs. Marian Keech, of
> 847 West School Street, says the prophecy is not
> her own. It is the purport of many messages she has
> received by automatic writing, she says. . . . The
> messages, according to Mrs. Keech, are sent to her
> by superior beings from a planet called "Clarion."
> These beings have been visiting the earth, she says,
> in what we call flying saucers. During their visits,
> she says they have observed fault lines in the earth's
> crust that foretoken the deluge. Mrs. Keech reports
> she was told the flood will spread to form an inland
> sea stretching from the Arctic Circle to the Gulf of
> Mexico. At the same time, she says a cataclysm will
> submerge the West Coast from Seattle Washington
> to Chile in South America.[7]

Now, your ordinary scientist might have stayed as far away from this as possible. I mean really—this is *National Enquirer* stuff, and potentially hazardous to one's career. Well, not Leon. He and a team went to Lake City forthwith, where Mrs. Keech received another message: On December 20, an extraterrestrial visitor was to appear at her house around midnight to escort her and her followers to a parked flying saucer and take them away from the flood, presumably to outer space.

Leon's prediction, assuming that the momentous event did not occur, was that the followers would attempt to reduce their dissonant state, produced by not having their beliefs confirmed, by attempting to convince others of those beliefs. A vast set of experimental data now supports that view, but at the time it was brand new. The clock struck twelve in Lake City. The group waited. No alien visitor arrived to take them to them to a flying saucer. An awkward period began among the believers waiting in Mrs. Keech's living room. But a few hours later, Mrs. Keech received another message:

> . . . *For this day it is established that there is but one God of Earth and He is in our midst, and from his hand thou hast written these words. And mighty is the word of God—and by his word have ye been saved—for from the mouth of death have ye been delivered and at no time has there been such a force loosed upon the Earth. Not since the beginning of time upon this Earth has there been such a force of Good and light as now floods this room and that which has been loosed within this room now floods the entire Earth. As thy God has spoken through the two who sit within these walls has he manifested that which he has given thee to do.*

Suddenly, the room was in better shape, and Mrs. Keech reached for the phone to call the press. She had never done this before, but now she felt that she must, and soon all the members of the group had called various branches of the news media. For days this sort of justification went on, and Leon's prediction was spellbindingly confirmed.

All of this came from a man who loved to play backgammon for hours on end. Unlike many academics, who talk a great deal without listening, Leon was a conversationalist. That back-and-forth reinforced his game behavior. Like all the truly great intellects I have known, he both listened and elicited testimony from the speaker. Being lectured to or lecturing himself turned him off. In the long run, what was so captivating for me was experiencing his insistence on avoiding the trap we all fall into: being transfixed by simple correlations and conclusions. He always dug below the surface.

The eloquence of cognitive dissonance theory surely guided me toward my ideas on brain modularity and belief formation, which I wrote about in *The Social Brain* many years later. Those priceless lunches and road trips to faraway places, including archaeological digs, all taught me about the richness of psychological concepts and provoked my thinking. My life was changed forever by knowing Leon. It was the scope and intellectual energy he brought to everything, from potato pancake recipes to complex mathematical formulations about brain scan data. His canvas had no boundaries, nor did his friendship. I had never experienced anything like it.

BACK AT THE LAB

Meanwhile, the work on motivation and reinforcement was moving forward. Premack had given me one of the experimental systems he'd built for testing motivation mechanisms. He liked an idea of mine that followed up on one of his theories and happily provided me with the gadget to test it. It was a device he'd rigged to allow a rat either to run in a wheel or to drink water. More specifically, the device was able to measure the responses of a rat when it did either. My question was, Would an adipsic rat (a rat that will not drink as the result of a specific brain lesion) drink if it was rewarded with an opportunity to run? Rats normally love to run, and if deprived of doing so, they will seek ways to run. If the adipsic rats would start to drink in order to have that opportunity to run, it would urge a more dynamic view of brain function and caution against the ever-growing tendency to see rigid one-to-one models relating structure to function. In fact, we learned that adipsic rats gladly drank if that was what they had to do in order to run (Figure 16).[8]

Again, in some sense we were seeing a cross-cueing strategy that allowed the brain, a dynamic, ever-changing system constantly in action, to switch strategies in order to accomplish a goal. In this case, the experimenter was creating new contingencies (if you do this, then you get that) that evoked new strategies. More generally, it revealed that it was probably always dangerous to claim that a particular brain network has a monopoly on any particular behavior.

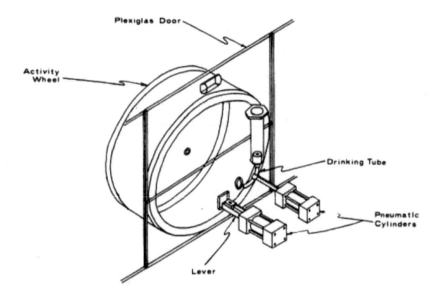

FIGURE 16. In order to get the activity wheel to spin for ten seconds, the rats had to take five sips of water. They were soon drinking like drunken sailors.

The brain is wily and does not follow simple rules. If one network is knocked out, a detour is rigged. It was a striking and, to my mind, an important finding, but, as it sometimes goes, it has been largely ignored.

Still, once hooked on the idea, I tried all kinds of experiments to further demonstrate that specific functional parts of the brain are very much a part of an inherently dynamic system. In one wild experiment, I tested monkeys with inferior temporal lobe lesions that rendered them unable to learn the difference between two visual patterns for a food reward. I wondered whether they would learn the new discriminations if they were given the opportunity to run in a large monkey wheel I had specially built for them. Here I discovered that I had more in common with monkeys than I'd thought. Monkeys hate to run in a wheel. Instead, a group of temporal-lobe-lesioned monkeys actually learned new discriminations in order to lock the wheel so it wouldn't move![9] Capitalizing on that preference for a stationary wheel, I found that visual capacity developed in monkeys even though the neural pathway that normally enables such visual learning was missing. Same point, different species.

FIRST STEPS INTO THE NEUROLOGIC CLINIC

I had finally contacted the New York University School of Medicine with the hope of studying neurologic patients with disorders such as global aphasia—patients who could not use or understand language due to a brain lesion of the left cerebral hemisphere. It is a devastating condition and it puzzled me. Was it that the remaining intact right hemisphere couldn't cover for the left hemisphere, or was it that the tests used to try to elicit language function were poorly designed? Both Leon's and Premack's psychological perspectives were pounding away in me.

Requesting to work with patients in New York was no simple matter. How? Where? Who says? When? These are the kinds of arrangements that require a planned process and tremendous staging. Once again, luck was with me. NYU School of Medicine had a storied neuropsychology* group, once led by Hans-Lukas Teuber, who had moved on to be the head of brain science at MIT. With him gone, I went in to meet the then-chair of neurology only to learn he was most definitely not interested in neuropsychology and claimed to have barely heard of Teuber. But, as I say, luck was with me, and someone put me in touch with NYU's Martha Taylor Sarno, their expert on aphasia, and a research project was born. By then I was learning that everything in New York comes with a story, and I remember one in particular.

It was Thanksgiving Day. I got a call to get up to NYU for a special testing of a patient. We had a car by then, so I drove up to the

---

* Neuropsychology studies the structure and function of the brain to understand how they relate to specific psychological processes and behaviors.

medical center on First Avenue. Now medical centers have long swaths of street parking reserved for M.D.s, who are easily identified in New York since they have their identity stamped on their license plates. On a normal business day, one would never park there unless authorized to do so by virtue of possessing one of these plates. Its being Thanksgiving, however, *no one* was using the spaces. I was in a hurry, parked my car in one of the normally coveted places, and went in to see the patient. All was well until I came out to drive home. There it was, a freshly minted parking ticket. I was furious because it made no sense to me. After all, I was doing medical research on a holiday for the good of humanity, or something like that. I chose to fight the ticket and sent off letters to the city, the whole bit.

About three weeks later, I got a call in my office at work. It was the New York City parking commissioner. He said, "Say, Doc, we got your letter and we agree with you. Your ticket has been waived, canceled, forgotten." I said, oh that's great, then he added, "Now, Doc, you teach at NYU, right?" Yeah, I said. He said, "So listen, my daughter goes there. I want you to do me a favor. I want her to be able to call you if she ever gets in trouble, okay?" Holy smokes! I was really in New York now. Everything and everybody is leveraged.

Back in Santa Barbara, Premack had moved on from studying motivation and instead was testing the mental structures of a chimp by teaching her crude communication skills. It too was ground-breaking research with immediate possible implications for patients suffering left hemisphere stroke. Again, our simple idea was that a stroke patient with a damaged left hemisphere had a whole intact right hemisphere that might be able, if trained correctly, to compensate for the left's damage in some way. David and I had wondered about this before I left for the East. Why couldn't we teach an aphasic the kind of metalanguage system his chimp had learned, which would allow for a crude kind of communication from the remaining right side of the brain? Maybe we could open up untapped lines of communication with these devastated, largely mute patients.

Of course, I knew all about what isolated right hemispheres

could do. We had worked for five years on getting right hemi-spheres of split-brain patients to do more than rule supreme on visual-motor tasks, such as the block design test. We knew that some of them could even read simple nouns. Scant data existed, however, on how well the right hemisphere could think symboli-cally. Premack's chimp work pushed us to try the idea out. If chimps could learn a simple symbolic system, then why couldn't a surviving right hemisphere?

What Premack had achieved with a chimp was inspired. He was able to show very elegantly that chimpanzees can do analogies. He reasoned that an animal can make a judgment of whether two objects are the same or different without understanding anything about the relationship between the two objects. He put it this way:

> *Sameness/difference is not a relation between objects (e.g., A same A, A different B) or properties, it is a relation between relations: For example: consider the relation between AA and BB, CD and EF on the one hand; and AA and CD on the other. AA and BB are both instances of same; the relation between them is "same." CD and EF are both instances [of] different; the relation between them therefore is "same." AA is an instance of same, and CD an instance of different; the relation between them is "different." This analysis set the stage for teaching chimpanzees the word "same" for AA, and "different" for CD. When taught these words, chimpan-zees spontaneously formed simple analogies between: physically similar relations (e.g., small circle is to large circle as small tri-angle is to large triangle), and functionally similar relations (e.g., key is to lock as can opener is to can.[10]*

Now that is an ingenious analysis. On the surface it seemed unlikely a patient with a major language disturbance could match the feats of a chimp. It was time to do experiments. Along with my new NYU graduate student, Andrea Velletri-Glass, and with Premack, still in California at UCSB, we began. With airplanes, phones, and fax, it somehow worked. We studied several patients intensely and discovered that patients with left hemisphere lesions, who were rendered severely aphasic, could nonetheless learn, to varying degrees, the artificial language that was successfully trained to chimps. In other words, we thought at the time that in some of the patients, the spared right hemisphere could reason at least to the level of a clever chimpanzee.[11]

The right hemisphere was, we assumed, carrying out the cognitive work in these experiments. Some of the patients were so severely damaged on the left it was the only logical possibility. At the same time, our belief that the right hemisphere should be able to do at least simple tasks was challenged by another study we carried out at NYU.[12] It deals with a phenomenon called "pure word deafness," the inability to comprehend spoken words. Patients with this problem could still read and understand words presented to them visually, but they could not understand words spoken to them. This was caused by a specific left hemisphere lesion. Okay, we said. But these same patients also have right hemispheres that are working. Why can't the right hemisphere take up the slack and understand spoken language? After all, the disconnected right hemisphere of some split-brain patients could understand spoken words. Remember, in those days we were still thinking in very simple ways about how the brain worked.

It wasn't long before a patient came along who allowed us to test the idea—once again proving the richness of the neuropsychologic clinic. The natural accidents of nature provide unending insights into how our brains are organized.

Case W.B. had been a business executive when he suffered a stroke that left him with his strange inabilities. While he could read perfectly well, write perfectly well, and had a fairly normal audiogram (hearing test), he could not understand spoken language. So, show him the word *knife* printed out on a card and he could say it, write it, and find the object from a grab bag of objects. On the other hand, if you said the word *knife* to him, he was unable to make any meaningful response. In test after test this condition persisted for the rest of his life. This simple finding was the exact opposite result from what we expected as a result of our split-brain work.[13] It got us thinking that maybe our thoughts about right hemisphere language were too general, too expansive. While there was no question that some language-understanding capability was present in the right hemispheres of N.G. and L.B., maybe they were the exception, not the rule. Maybe the positive finding that global aphasics were able to carry out simple chimp-level analogies was illusory as well. Maybe it was remnant parts of the left hemisphere carrying out that deeply cognitive task. Maybe, after all, most people only possessed language in their dominant hemisphere. Fifty years later, the field is still struggling with these questions.

CHALLENGING THE IDEA OF TWO MINDS

The tug of the new intellectual issues pushed by Premack and the weekly cross-examination about everything from Leon found me constantly reassessing the earlier claims Sperry and I had made about two minds in one brain. Added to this constant discussion were frequent trips to New York City by Donald MacKay, the sophisticated neuroscientist/philosopher whom I had brought to UCSB for a few months in 1967, and who also had issues with the first split-brain story of two minds. In tolerating the idea of two minds, Premack and Festinger gave moral support; meanwhile, MacKay simply didn't buy it. Over martinis and Manhattans at Il Bambino, we tried to hash things out. Even MacKay, a physicist and a practicing Presbyterian, would throw back a Manhattan as we waited for the shrimp scampi to arrive.

Out of the blue, a request came in to write a review of split-brain research for the *American Scientist*, the journal of the Sigma XI organization. It was a good place to discuss such a grand issue as "one brain, two minds?," which in fact became the title of my article. As I reread the article now, some forty years after I wrote it, I can see the tug-of-war that was starting in me. While I staunchly defended and argued for the double mind idea, I called upon the new kind of experiments and data I had developed at NYU. These would, in the long run, find me changing my views and facing a system much more complicated than the one I had defended.

In the paper, I started out with a summarizing statement that the right hemisphere "could read, remember, write, emote and act all by itself. It can do almost anything the left can do, with admitted limitations in the degree of competence."[14] I went on to say that while some of us focused on this kind of result, others focused on what may still be connected and transferring through lower brain systems, or how each hemisphere might have different cognitive styles of handling the sensory information it received.

It was to all of this, no matter what the current researchers in split-brain work were doing, that MacKay raised the idea of "normative" systems. This old chestnut from philosophy, worked hard by one of MacKay's countryman, David Hume, stated that beings like humans have certain behaviors and thoughts that are part of the human condition: All that we do is normative, that is, concerned with following the directives of those core preferences and capacities, even though they may be culturally learned.[15] MacKay was arguing that this is what people do and that no internal disconnection can change our normative stance on the actions we take. When these philosophical ideas were applied to brains, his views were consistent with the thought straight out of Wikipedia, "Normative statements and norms, as well as their meanings, are an integral part of human life. They are fundamental for prioritizing goals and organizing and planning thought, belief, emotion and action and are the basis of much ethical and political discourse."[16]

This sort of thinking and framing of the issue of split brains was novel and seemingly distant from actual experimental studies. But MacKay kept hammering at it with questions: How can each hemisphere have two different prioritizing systems, two different evaluations of a common stimulus? How can one like an orange and the other not like an orange? It was definitely time, again in the Francis Bacon tradition, to leave Il Bambino and get into the lab and see what could be done.

In particular, MacKay wanted to see more direct evidence. He wanted to see both hemispheres ready to act but with each holding different evaluations of the same stimulus. He wanted to see the right hemisphere love PB&Js and the left hate them and the two duke it out for lunch. We first started to get at this with monkey research.

Alan Gibson, a graduate student at NYU who had come with me from UCSB, had the bright idea to make a lesion on one side of the hypothalamus of a monkey. The hypothalamus, which is at the base of the brain, controls much of our eating behavior. What if half of it were damaged? Would a split-brain monkey then be less motivated to act for a food reward when viewing the world through the hemisphere associated with the lesion? Would it act normally when using the other hemisphere? Both hypotheses turned out to be true.[17] Each hemisphere seemed to possess its own preference and prioritizing system. Mounting evidence for separate normative systems, but still not enough. We kept after it.

J. D. Johnson, also a graduate student, and I did another experiment on monkeys. I must say this one was pretty clever. Split-brain monkeys were trained to learn a simple visual discrimination through one eye on what is called a fixed reward schedule, or FR 2. This means that on every other learning trial, the half brain viewing the visual task is rewarded if the response is correct. So, rewards come every second trial, not on every trial. Still, the half brain learns the task well, no problem.

Now comes the fun part. While the hemisphere that has learned the problem performs the task, the naïve hemisphere is allowed to view the behavior of the trained hemisphere, but only on the trials where no reward is present. We had already done studies where the naïve hemisphere watched the trained hemisphere when rewards were present on every trial. Under those conditions, the naïve hemisphere learned quickly. But what if the naïve hemisphere watched the other hemisphere perform correctly but without reward? Could it learn under these conditions, too? After all, if the normative system was operating and pervasive, both hemispheres should be tuned in to the fact that the stimulus choice associated with the trained hemisphere's consistent correct response was tied to a nice reward.

Again the results were striking. Not only was there no suggestion that the naïve hemisphere knew the positive value of correct stimulus; the naïve half brain tried to disrupt the choices of the trained brain.[18] It was like another animal fighting over food. What was also emerging was the difference between information playing into normative processes that could remain what we called "cold," versus information that was "hot," that is, emotionally laden.

The best example of this had come a few years earlier when we tested Case N.G. in Santa Barbara. In that test, the goal was for the left and right hemispheres to learn, without being told, to choose, say, the numeral "1" when presented with the choice of a "0" and a "1." In one stage of the experiment, we gave reinforcement only to the left hemisphere by flashing the word *right* or *wrong* after the correct or incorrect response. When we did this, the patient's left hemisphere learned the task quickly, but because the feedback had only been given to the left brain and it did not leak over in some way, the right hemisphere did not learn. In short, it appeared the right hemisphere never got a cue from the left hemisphere about which light to continue to push, so it chose randomly.

In a second stage of the experiment, I admonished the patient for making such a simple mistake (when the right hemisphere was wrong). She blushed and was embarrassed. Emotions are generated from parts of the brain that have not been separated and thus both hemispheres are privy to them. The emotion of embarrassment produced by my admonishments now served as a feedback cue, a negative feedback cue, to the right hemisphere. From that point on, the right hemisphere quickly learned the task in normal time. What was so remarkable, however, is that what the right hemisphere was learning was not what the left hemisphere thought the right hemisphere was learning. Thus, when I asked her, following all of this, how she made her choice, she (her left hemisphere) said she picked the "1." What her right hemisphere actually had learned instead, however, was not to pick the "0." Again, her left hemisphere didn't know what the right had learned, and the right had learned because it had been cued by embarrassment during the training phase.[19]

NOT ONLY SMART BUT A SMARTY-PANTS

Living in New York provided all kinds of new reinforcements for me, too. My contact with Buckley increased, and he invited me to several of his *National Review* editorial dinner parties where his senior editors would blow off steam after a hard day of work putting the current issue of the magazine to bed. The Buckleys hosted social occasions several times a week, which for mere mortals would be in the category of "too much." Most people have thrown dinner parties at some point and have most likely come to the conclusion that they are enjoyable at first, but almost invariably drag on too late, since no one quite knows how to end them. As a result, it usually takes several months for another dinner party to seem like a good idea.

The Buckleys solved the dinner party exit dilemma with elegant precision: The guests arrived at 7:45 and had drinks until 8:15. Dinner was then served until about 9:20, whereupon the company repaired to the living room for coffee and cigars until 9:50. At that point, some confederate in the group made it clear to one and all that it was time to leave. By 10:00, the party was over, Bill went upstairs to write his column for the next day, and everyone was happy. I have adopted the Buckley method to great advantage, although I don't use the confederate angle. At 9:30 (I am not the night owl that Buckley was), I just tell everybody that it is time to go. Charlotte and I have made dinner parties a big part of our life, and I would venture to say that over the past thirty-seven years, the three hundred or so we have given have played a consequential role in contributing to the field of cognitive neuroscience.

One night in 1971, the schedule at the Buckley's was thrown off. Daniel Ellsberg had just leaked the Pentagon Papers, the Edward Snowden–scale leak of the day, and Bill and his feisty editors came up with the idea of publishing a spoof on the papers. Since I was sitting in the room, I was assigned to write some memos from Dean Rusk, who had served as secretary of state from 1961 to 1969, about the Vietnam War. A few years earlier, Bill had rejected and sent back a manuscript that I had submitted for publication in the *National Review* on the Watts riots with the comment "Return to Mike Gazzaniga in a plain manila envelope" penned at the top. I am sure he was wary about the prospect of my pulling off this assignment, but it turned out to be easy. Speaking in governmentese in memo form was a snap. Years of writing grants and memos in large universities, where the very life of language is squelched, had admirably prepared me to pound out why Rusk thought the war should be ended quickly. All government systems develop acronyms, and so I came up with STW, which meant Short Term Warfare. The alternative, LTW (you got it, Long Term Warfare), was not acceptable to the American people. And on and on.

Bill loved it, and a few days later, the story hit. Walter Cronkite didn't realize it was a hoax and opened the *CBS Evening News* with a picture of the cover of the *National Review* issue about the Pentagon Papers, and calls poured in to the magazine. Somebody called the retired Dean Rusk in Atlanta and read him my memo. He said that although he didn't recall the memo, he might have written it. I am still not quite sure what Bill had in mind with this whole plan. He was off in Vancouver for two or three days, unavailable for comment. When he arrived home a couple of days later, there was a

FIGURE 17. My entrepreneurial gene was at it again. I talked Bill Buckley into interviewing my friends on TV, which he did with gusto. The top photograph shows Buckley questioning (*left to right*) B. F. Skinner and Donald MacKay. In the second photograph (*left to right*), Nathan Azrin and David Premack are in the hot seat. In the lower photograph, Skinner (*center*) speaks with Leon Festinger (*right*).

huge news conference, and in one of the great verbal dodges of all time, he came up with some profound moral reason why he had perpetrated the stunt. I was convinced at that moment that Bill should be elected president of the United States for the sheer fun it would bring. William Randolph Hearst Jr. called the prank "one of the most sensationally successful spoofs in the history of American journalism."

On a different front, I had convinced Bill he ought to have great mind scientists on his TV show, *Firing Line*. I mean, how many politicians can you listen to? So he did. First off was a show featuring the great behaviorist B. F. Skinner and my pal Donald MacKay. It was on the nature of personal freedom. Calling it high level would be shooting low even by the usual Buckley standard of erudition. For years, it would be his most popular show. A couple of years later, I convinced Bill to do a series of shows. He got Leon to do one again with Skinner, on the mechanisms of moral development. In another show, Premack and Nathan Azrin, a psychologist who thought anything could be trained into anybody, discussed the limits of behavioral control. I was quite proud to have played a role in getting this kind of discussion on television (Figure 17). More broadly, it showed once again that real cultural leaders, no matter their politics or their background, can play together just fine. That discovery has had a huge influence on my life.

ON THE MOVE AGAIN

The family was growing. With three young, energetic daughters, it was time to try the suburbs. We chose Weston, Connecticut, for a variety of reasons, one of which was its pastoral beauty. Still, it was a two-hour commute each way, each day, which meant four hours of suspended thought. The mornings were fine and, in fact, enjoyable. Hit the station, grab a coffee and the *Times,* sit back in a comfortable train, and off you go to Grand Central Terminal and a subway ride down to Greenwich Village. The energy level was high, and since everybody else was doing it, it all seemed normal. Coming home at night was another matter.

Fatigue was what got me. The end of the day called for a beer, a copy of the *New York Post,* and a hope for a seat on the train to Westport, my stop. The train floor becomes sticky with the commuters' beer, and the belligerence level goes up. All in all, it was not a whole lot different than a German beer hall. Just a few years of this grind did me in.

Out of the blue, I got a call from the State University of New York, Stony Brook, and was asked if I wanted to move there. I immediately said it sounded interesting and went out for the usual job talk and dinner. I liked it all. It too had an excellent department, and was in a beautiful location that would require no commuting. We moved to Stony Brook, on Long Island about sixty miles from Manhattan, that summer.

Just before I left for Long Island and my new life, I got another call, this one from Dr. Ernest Sachs, up at Dartmouth Medical School. He was head of neurology at the time, and he invited me up to give a lecture. I was thrilled. I was to play the role of professor at my old alma mater! It was especially sweet because the very same

FIGURE 18. Donald Wilson (*left*) and David W. Roberts, Wilson's resident at the time, launched the Dartmouth series of split-brain patients. Roberts went on to become the head of neurosurgery at Dartmouth and the inventor of a computer-based operating microscope.

medical school had rejected my application eleven years earlier, even though I was an undergraduate at Dartmouth and my brother was one of their stellar graduates. It is events like this in one's past that fall off the story line. What if I had been accepted and gone? There would have been no split-brain work for me. How would that whole story have been different? I believe that things just happen in life, and pretty much after the fact, we make up a story to make it all seem rational. We all like simple stories that suggest a causal chain to life's events. Yet randomness is ever present.

Of course, even more important when we choose a new course in our lives are the new people we meet as a result. Stony Brook proved a rich experience for me, both scientifically and personally. I was lucky to have a series of outstanding graduate students, in particular Joseph LeDoux, Mr. Creativity and Energy personified. After receiving his Ph.D. with me, he went on almost single-handedly to put the field of the neuroscience of emotion on solid footing. Joe, from southern Louisiana, is a musical Cajun (is that redundant?) at heart and at night grabs his Stratocaster to play with his band, the Amygdaloids. Never far from neuroscience, their CDs are titled *Heavy Mental, Theory of My Mind,* and *All in Our Minds.* Had I not taken the job, I might never have had the opportunity to know him.

At any rate, after the lecture at Dartmouth, a young neurosurgeon, Donald Wilson, approached to say he had sectioned the callosums of some patients and would I be interested in studying them (Figure 18). Would I ever! Wilson had started a new series of cases at Dartmouth but nobody was working with them. He too had decided the surgery could help those patients who were not being successfully managed with anti-epileptic drugs. In the California series, both the anterior commissure, which resides deep in the brain, and the corpus callosum were sectioned. He felt the surgery could be improved, in terms of outcomes, if sectioning the small anterior commissure (a small bundle of nerve fibers that, similar to the corpus callosum, joins parts of the two hemispheres) could be avoided. In cutting the anterior commissure, one had to enter structures called the lateral ventricles, a process that sometimes introduced infection.

Wilson also introduced another new technique. Cutting the entire corpus callosum was a long procedure, almost seven hours. He thought it would be less traumatic on the patient to do the surgery in two stages. Thus he cut the posterior half of the callosum and then a few weeks later the anterior half. As I'll explain, this allowed for some major insights into callosal organization.

I could barely contain myself. I had desperately missed studying split-brain patients and was eager. First, I had to figure out how and where to test the several patients. It soon became clear that I needed to test them in their New England homes, which were spread all over Vermont and New Hampshire. How was that going to work? To get going, I simply decided I would haul testing equipment into their homes and do it like I had done it in Los Angeles. That proved to be short-lived. While there were some notable exceptions, many of the patients lived in remote trailers that did not lend themselves to this sort of thing. Then the idea of a trailer was born. I went back home and I bought a Del Rey trailer that I could haul behind the family car. If I remember correctly, it cost fourteen hundred dollars, and a neighbor and I converted it into a lab! Now my mobile lab could be driven anywhere, and we could study the patients in our professional space, leaving the patients' families to their own private space. Our mobile lab didn't get its upgrade until years later.

By the time we actually had pulled up stakes in Connecticut and moved to Long Island, the new split-brain testing program was launched. Multiple trips to New England slowly established the fact that a growing and important population of patients was becoming available for testing. There were, nonetheless, major logistical problems. Driving up to New England from Weston was a relatively short affair, while driving from Stony Brook, either indirectly by driving toward New York City and the Throgs Neck Bridge, or by taking the Port Jefferson Ferry, was a challenge. We had traded the family car for an orange van, freshly driven in from California, following a cross-country trip and summer vacation, and it proved to be a lifesaver on more than one icy road trip (Figure 19).

The real advance in the research program, however, came not from the testing gear and trailer, but from my bevy of new graduate students. Energy and smarts is what counts, and they all had it. Our frequent road trips to New England became legendary, and of course they were fun. When new postdoctoral fellows would come to be interviewed for jobs, they were intent on reviewing what they had accomplished in their thesis work. I would look at them blankly as

FIGURE 19. The family van/trailer at work in Brattleboro, Vermont. The big orange van hauled around the little trailer to its left. Our trailer plugged into the trailer that Case P.S. lived in for years (*far left*).

they finished, because, even though they all had accomplished something of value, I had a critical question: "Do you drive?"

DON'T QUIT YOUR DAY JOB

The penal colony architecture of the State University of New York, Stony Brook, was an anomaly in the otherwise idyllic setting of the Long Island shore. Stony Brook, Setauket, and Port Jefferson, snuggled into the north shore, were roughhewn and breathtaking. Governor Nelson Rockefeller decided to compete with the University of California system in the early 1960s by building Stony Brook, but something went wrong in the design department. It was as if the designers had never left Albany to look at the gorgeous setting and see its aesthetic potential. For years, articles were written about the depressing nature of the campus and reported all over scientific journals.

None of the physical shortcomings of the university, however, seemed to thwart their hiring of a terrific faculty. By the time I joined the university, it had been in business for only eleven years. Even though New York State had imposed its awful bureaucracy from the start, the campus's energetic faculty made it feel like a Silicon Valley start-up. Collaborations were easy to form across disciplinary lines, and I was beginning to think about combining biochemical approaches to the study of learning by using split-brain pigeons. Maybe Stony Brook would be a place to do that, too.

As I said, energetic graduate students are the key to so many adventures in science. Eager and committed, they find you. If they're smart, stuff really begins to happen. They are the legs, the energy, and the future for any science, and Stony Brook had more than its fair share. One of them, Nicholas Brecha, showed an interest in the wacky pigeon idea and is responsible for getting it off the ground.[20] The project required learning a sophisticated behavioral training method, anatomy and surgery, and, of course, biochemistry. Before you knew it, Brecha had mastered all three by calling upon campus specialists, all of them scientists in their own right. After a few years of hard work, the project was completed, and, unfortunately, we couldn't find any differences between the trained and untrained half brains of a pigeon. The bet was placed, the work was done, and bupkis. That is the way it goes and why one always has side projects. Brecha, for example, went on to a successful career and is now an expert on the retina and a professor of medicine at the University of California, Los Angeles.

Many of my other projects were outside straight academic work. Ever since my Sol Hurok days as political entrepreneur at Caltech and my failed attempt to start a new audiotape company with Bill Buckley (another story), I was stuck with a compulsion to be unconventional.

Steve Allen Jr., the son of my late comedian friend, is a physician. We became quite close, and he encouraged the wacky idea I had to make science documentaries. He is hysterically funny, humane, and, like his father, captivated by brain research. At one point, we cooked up an idea to make a film about the brain and creativity. After my move to Stony Brook, I had the Beaulieu 16 movie camera that allowed for sound to be recorded right on the film during filming. This was an improvement on the old Bolex 16 that I had used at Caltech, with its double-sprocket film and its awkward process for adding sound. I thought the Beaulieu would make editing and production easy. While the camera was bought to use for my patient work, I thought it could double as an aid to the noble goal of scientific education.

The camera, parabolic sound microphones, lights, and all the rest required several bags. All of them were heavy or awkward to carry by myself. Undaunted, I called Steve Sr. and asked if I could interview him about creativity. He couldn't have been more agreeable. Off to Los Angeles I went, carrying film gear through airports with a certain swagger.

I arrived at his home one Saturday morning, and Steve was still lounging around in his blue robe. I didn't think people actually lounged in their robes except in the movies. He walked me into his living room and suggested ways for me to set up the lighting, tripods, and all the rest. It all started to become surreal, and this little voice in me said, What are you doing? Why are you bothering this guy in his robe? Why aren't you back at Stony Brook doing your research? Who do you think you are—Fellini? I was just about to leave, making some excuse, when Steve said, "Looks like you are ready." With that he started to play one of his own compositions, "This Could Be the Start of Something Big," and for a moment, I did think I was Fellini. The experience was exhilarating, and I vowed to take my gear everywhere to capture moments for the film. In fact, I took the gear to Paris a bit later, set it up in my hotel room at the Paris Hilton, and threw open the window. With the camera on automatic, I filmed myself standing in front of the window, with the Eiffel Tower in the background, and thought my second career was launched.

Oh my, we do crazy things. After filming a ride up the Eiffel Tower and a tour around half of Paris, I went home, loaded up with my footage. With great anticipation I waited for it to come back from being developed, stuck it in my projector, and sat back to savor my ingenuity. Let me simply say my escapade into filmmaking ended abruptly. My favorite disastrous scene was the one in the hotel window. Because my camera was reading the light level of the bright Paris sky, the guy in the foreground looks like he is in a witness protection program. Of course, one always looks for the silver lining. Steve looked terrific in his robe.

NEW PATIENTS, NEW DISCOVERIES, NEW INSIGHTS

Meanwhile, the research team was forming to tackle the new human split-brain work. Gail Risse led the charge, and others were soon to join. I also brought my monkey program out from NYU and loaded it up with new students, such as Richard Nakamura, who years later became the deputy director of the National Institute of Mental Health. We were busily exploring the question of whether one brain was as good as two. Cigar smoking and gentle, Richard preferred to stay with the monkey work. Meanwhile, Joe LeDoux was losing interest in his animal project, and I recruited him into the human work.

The split-brain team worked hard but at first the results were thin. The first patients were a complicated group. While the neurosurgical reports suggested that most of them had had a full callosal section—reports we accepted at face value—it was evident, upon much neuropsychological testing, that the surgery had not been complete. Before we knew the reports were flawed, we thought we had discovered an interesting fact. Unlike the California patients, who had both the callosum and anterior commissure sectioned, the new patients explicitly had the smaller anterior commissure intact. If any kind of information transferred between the hemispheres in these patients, we, not knowing that their callosal sections were incomplete, assumed it would be due to the intact anterior commissure. We knew that an intact anterior commissure in a monkey allowed for all kinds of visual information to transfer.[21]

In the end we got it right, but we definitely went through a phase where we got it wrong, which became evident when we combined months of neuropsychological testing with newly emerging EEG data from the Dartmouth neurologists.[22] At first we thought we had seen evidence for transfer of visual, somatosensory, and auditory information between the hemispheres and concluded the anterior commissure was the source of that cross-integration. We began to think animals and humans were more alike than not on this parameter.

It turns out that the first group of patients had variations in their partial disconnections. Some were partial splits by design. For example, a case might undergo anterior callosal surgery first. If the seizures came under control, then no further surgery was carried out. In other cases, though, there was inadvertent sparing of the anterior callosum. For example, in one case the anterior callosum was sectioned, and months later the posterior callosum was sectioned. The surgeon, however, had inadvertently left some anterior callosal fibers where the two surgical sections were to intersect. At the time, neither we nor the surgeon knew this. This patient showed transfer. We assumed it was due to the uncut anterior commissure, since we all assumed the first surgical reports were correct and a complete section of the corpus callosum had been accomplished. A few years later, the EEG results illuminated the story.

These roller-coaster results were no fun. We were beginning to back off from testing our New England patients when it all changed and we began to learn some things. The parts of the callosum that were sectioned did produce some specific modality deficits in interhemispheric integration. That is, specific areas of the callosum integrate specific kinds of sensory information such as vision and touch.[23] But it was indirect evidence and it wasn't clean. We were all beginning to think we should begin other avenues of research.

Then along came Case P.S., a teenager from Vermont, who led us out of our confusion and revived our interest. P.S was reported to have had his entire callosum sectioned in one operation by the Dartmouth surgeons. Even though the Dartmouth procedure required leaving the anterior commissure intact, he was "split," for sure. In a matter of weeks it was clear as a bell that a truly fully sectioned callosal patient with the anterior commissure unsectioned was identical to the Caltech patients in terms of the disconnection effects. Nothing transferred between the hemispheres; each hemisphere seemed specialized in its own way. The trips to Vermont became monthly and stayed that way for many, many years.

Many things were immediately evident upon testing P.S. There was, flat out, no interhemispheric transfer of visual information. Visual stimuli presented to the right hemisphere stayed isolated to that hemisphere and could not be named or described by the left

FIGURE 20. Joseph LeDoux was one of the first scientists to work with a viable series of split-brain patients out of Dartmouth. Today he is considered one of the neuroscientists primarily responsible for the scientific study of emotion. On the left is the GMC van that Joseph convinced the National Science Foundation to buy us so we could continue our work. (LeDoux photograph by New York University photo bureau.)

hemisphere. This meant the anterior commissure did not transfer visual information as it had in the patients who still had some uncut callosal fibers. P.S.'s tests offered evidence that the human brain was organized differently than a monkey's brain; a fully callosal-sectioned monkey with an intact anterior commissure could transfer visual information between the hemispheres. Of course, it also meant the Dartmouth, or East Coast cases, as they were to be called, were just like the California cases. This fact would prove to be a sore point between the two research groups in the years ahead. In ideal science, replication is key and a virtue, and everyone warmly collaborates. But science conducted by mere mortals often falls short of this ideal.

LeDoux was amped (Figure 20). His introduction to the so-called split-brain patients had been the earlier patients in the Dartmouth series, and while of interest, they were not compelling. Case P.S. was loaded with phenomena, and LeDoux captured many of them. He knew the earlier scientific literature cold and would say, "Let's try this," which might be to ask the patient to draw a cube with the left or right hand. After experiencing months of confusing responses that

FIGURE 21. Case P.S., the case that put us back on track. He was a warm and affable teenager. On one of our trips to California to study his brain waves, we took him to Disneyland. Here he is pictured on the trip with his mother.

had come out of the first group of patients, his jaw dropped when he saw P.S. easily draw a cube with his left hand but not be able to do so with the right hand. Back in our sparse motel room that night, I can remember LeDoux saying, "We finally have ourselves a split-brain patient to study."

In trip after trip, the dynamic nature of P.S.'s postsurgical course revealed itself (Figure 21). Unlike Case W.J., the arm ipsilateral to a particular hemisphere quickly came under the control of that hemisphere. Again, that meant either hemisphere could come to control not only the contralateral arm, but also the ipsilateral arm. And that meant drawing a cube correctly could soon be accomplished by both arms/hands.[24] LeDoux logged all of these changes, and within fifteen months, both hands were equal in skill. This learned control of both arms was evident in the other California patients, so it was not surprising. Still, it is exciting to see things unfold like they are supposed to.

P.S. was unique in so many ways, not the least of which was his spunky right hemisphere (Video 7). Very soon after his surgery, the right hemisphere, while unable to speak, was very responsive when nonverbal outlets were available to him. He was the first split-brain patient to respond to verbal commands to the right hemisphere, in addition to simple nouns. If a noun, such as the word *apple*, was flashed to the right hemisphere and he was asked to point to a picture that matched the word from among a set of pictures, P.S., like other split-brain patients, had no problem. Yet, unlike other patients, when a simple printed command was given to the right hemisphere, like "get up" or "point," he could do that, too. The right hemisphere didn't sit there like a lump on a log; it did stuff (Video 8). In fact, as we were soon to discover, it could have its own preferences. Having a more engaged right hemisphere to work with opened up all kinds of issues and studies (Video 9). LeDoux describes everything better than most people, especially me. He was my partner in all of these studies:

> Patient P.S. was especially important. He could use both sides of his brain to read but only the left hemisphere to speak. Previously, the right hemisphere had been thought of as a lesser partner, with cognitive capacities like a monkey's or chimp's, but not like a human's. The left hemisphere clearly had self-awareness, but whether high-level consciousness was possible on the other side as well seemed dubious. With P.S. we were able to ask whether the right side was self-aware because his right hemisphere could read. So we flashed questions to his right hemisphere and his left hand would reach out and, using Scrabble tiles, spell the answers. In these simple tests we found out that P.S.'s right hemisphere had a sense of self (he knew his name) and had a sense of the future (he had an occupational goal), both important qualities of conscious awareness. It was particularly interesting that the right and left hemispheres had different goals for the future. Might there indeed be two people in one head?
>
> In the process of testing the interactions between the two sides, one day in our camper trailer lab, Mike made an important observation. We were giving the right hemisphere written commands (stand, wave, laugh), and P.S. responded appropriately in each case. Had Mike not been there that's probably as far as it would have gone. We would have been happy to have shown that the right hemisphere could respond to verbal commands. But Mike's incredibly fast and creative mind immediately realized there was more to it. He started asking P.S. why he was doing what he was

doing. Remember, only the left hemisphere could talk. So when the command to the right hemisphere was "stand," P.S. would explain his action by saying he needed to stretch. When it was "wave," he said he thought he saw a friend. When it was "laugh," he said we were funny.

That was the birth of Mike's theory of consciousness as an interpreter: a reason for doing these things was made up to justify the impulse to take a certain action. This led to more experiments to directly test the idea.

On the next trip we simultaneously presented different pictures to the two hemispheres and told him to point to the card that matched the pictures. In the classic example, we presented a snow scene to the right hemisphere and a chicken claw to the left. The left hand pointed to a card picturing a chicken and the right hand to a card picturing a shovel. P.S. explained his choices saying he saw a chicken claw so he picked the chicken, and you need a shovel to clean out chicken shit in the shed. The left hemisphere, in other words, used his behavioral responses as the raw data to concoct an interpretation that was then accepted as the explanation of why he did what he did.

For the left hemisphere of a split-brain patient, everything done by the right hemisphere is an unconscious act. Mike proposed that our behaviors are controlled by systems that function unconsciously, and that a key function of consciousness is to make sense [of] (interpret) our behavior. This was his theory of the interpreter. . . .[25]

Joseph's flattering retelling of those trailer days fails to fully capture his role in the discovery. When something happens in a setting like that, everybody is equally involved. It's mutual cueing all the way and our only chore is to make sure the patient is not in on it. He wasn't.

In a sense, the insight that P.S. provided us (that the left hemisphere would come up with an explanation that made sense of the behaviors initiated by the right brain) came from changing our mind-set, not his. For the previous twenty years, split-brain researchers were intent on seeing what a particular hemisphere could do and could not do and whether there was information transferred between the hemispheres. This led us to ask a certain kind of question in a certain way. After we presented a stimulus to one hemisphere or the other we would ask, "What did you see?" It wasn't until twenty years later that we finally wondered, "What does the left speaking hemi-sphere think about all these things the right hemisphere is doing?" After all, the left hemisphere has no clue why the behaviors are hap-pening. Finally, it dawned on us in that cold trailer. Joseph and I asked, "Why did you do what you just did?" In simply changing the question asked of the patient, a virtual torrent of new information and insight flowed. Though the left hemisphere had no clue, it would not be satisfied to state it did not know. It would guess, prevaricate, rationalize, and look for a cause and effect, but it would always come up with an answer that fit the circumstances. In my opinion, it is the most stunning result from split-brain research.

Over the next few years, we hammered away at it with the patients we studied (Video 10), and the "interpreter" revealed itself in many classic experiments. The one described above was a typi-cal example of the speaking left hemisphere piping up with some kind of story to explain the actions that were initiated by the right hemisphere without the knowledge of the left hemisphere. At other times the left hemisphere would explain away emotional feelings caused by the right hemisphere's experiences. As I have mentioned earlier, emotional states appear to transfer between the hemispheres subcortically, and this transfer is not affected by severing the cor-pus callosum. Thus, even though all of the perceptions and experi-ences leading up to that emotional state may be isolated to the right hemisphere, both hemispheres will feel the emotion. Though the left hemisphere will have no clue why or where the emotion came from, it will always try to explain it away. For example, I showed a scary fire safety video about a guy getting pushed into a fire to the right hemisphere of V.P. When asked what she saw, she said: "I don't really know what I saw. I think just a white flash." But when asked if it made her feel any emotion, she said: "I don't really know why, but I'm kind of scared. I feel jumpy, I think maybe I don't like this room, or maybe it's you, you're getting me nervous." She then turned to one of the research assistants and said, "I know I like Dr. Gazzaniga, but right now I'm scared of him for some reason." The left hemisphere felt the negative valence of the emotion but had no knowledge of what the cause was. The interesting thing is that lack of knowledge does not stop it from coming up with a "makes sense" explanation that fits the circumstances: I was standing there and she was upset. Her interpreter put the two together into a cause-and-effect conclusion. I must have scared her.

The interpreter can affect many cognitive processes, including memory. For instance, Elizabeth Phelps, then a postdoctoral fellow and now a distinguished cognitive neuroscientist at NYU, and I showed a series of photographs to split-brain patients. The photographs told the story of a man getting up in the morning and getting ready for work. Later, we showed them another series of photographs and asked which ones they recognized. These photographs included the same pictures, some new ones that were unrelated to the story line, and some that were closely related to the story line. While the right and left hemispheres both accurately identified the previously seen photographs, the left hemisphere also falsely recognized the new pictures that were related to the story.

The left hemisphere has a tendency to grasp the gist of a situation, make an inference that fits in well with the general schema of the event, and toss out anything that does not. This elaboration has a deleterious effect on accuracy but usually makes it easier to process new information. The right hemisphere does not do this. It is totally truthful and only identifies the original pictures.

The interpreter also will explain away input from the body as illustrated in the following classic experiment (which most likely would not pass muster today with a human subjects committee).

Stanley Schachter and Jerry Singer told volunteers for an experiment that they were getting a vitamin injection to see if it had any effect on the visual system. What they really received was an injection of epinephrine and what the researchers really wondered was if the appraisal of their physical reaction would depend on the surroundings. Some of the subjects were told that the vitamin injection would cause side effects such as palpitations, tremors, and flushing, and some were told that there were no side effects. After the injection of epinephrine (which does have the side effect of palpitations, tremors, and flushing), a confederate of the researcher came into the room with the volunteer and behaved in either a euphoric or an angry manner. The volunteers who were told about the "side effects" of the injection attributed their symptoms to the drug. Those who were not informed, however, attributed their autonomic arousal to the environment. Those who were with the euphoric confederate reported being elated, and those with the angry confederate reported being angry. Three different, reasonable "cause-and-effect" explanations for the same physical symptoms came spewing out of their left hemisphere interpreters. Only one was correct, however: the injection of epinephrine.

So when it comes to the interpreter, facts are cool, but not necessary. The left brain uses whatever is at hand and ad libs the rest. The first makes-sense explanation will do. It looks for cause and effect and creates order out of the chaos of inputs that have been presented to it by all the other processes spewing out information. This is what our brain does all day long: It takes input from various areas of our brain and from the environment and synthesizes it into a story that makes sense.

THE RIGHT HEMISPHERE LEARNS TO SPEAK

Case P.S. was also the first to reveal another major reality in split-brain research—his right hemisphere actually began to speak simple

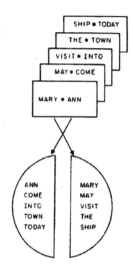

	RESPONSE
P.S. :	Ann come into town today.
E. :	Anything else ?
P.S. :	On a ship.
E. :	Who ?
P.S. :	Ma.
E. :	What else ?
P.S. :	To visit.
E. :	What else ?
P.S. :	To see Mary Ann.
E. :	Now repeat the whole story.
P.S. :	Ma ought to come into town today to visit Mary Ann on the boat.

FIGURE 22. Case P.S. began to make one-word utterances out of the disconnected right hemisphere, so we presented him with what we called the "triple story" test. We presented five word pairs in a sequence that would tell the following story: "Mary Ann May Come Visit Into The Town Ship Today." This is what our normal subjects recited when we presented them with the test. Each of P.S.'s hemispheres, however, discerned a different story line. The left brain saw "Ann Come Into Town Today" and the right brain saw "Mary May Visit The Ship." The dialogue above is how Paul's two hemispheres reported the experience. The left brain responded, and then the right brain suggested a different arrangement of words, leading to the summary response at the end, combining both outputs.

utterances. Soon after his surgery, he behaved like many split-brain patients. The left hemisphere could understand language and could speak. The right hemisphere could also understand simple language but could not speak. That was the standard situation. P.S., however, was beginning to behave differently. He startled us by uttering single words out of his right hemisphere.[26]

We knew it was right hemisphere speech because of a simple test. We would flash a picture of an object to each hemisphere and ask P.S. to name what he saw. About two years after his surgery, he began to name objects no matter which hemisphere saw them. To test to see if information was somehow transferring between the two hemispheres, we changed the question. We didn't ask "What did you see?" but "Are the pictures the same or different?" He couldn't do it. It was weird. If the pictures flashed separately to the right and left hemispheres had been, say, of an apple and duck, he could say "apple" and "duck," but these separated hemispheres couldn't compare what they saw and indicate if the pictures were the same or different. Of course, if both pictures had been presented to one hemisphere or the other, to say "same" or "different" was trivially easy.

We pushed hard on this. The ability of a right hemisphere to change and be able to speak was there, no doubt about it. Over the following years, this ability began to appear in other patients as well. We would later find that both Case V.P. and J.W. learned to speak out of the right hemisphere. In one test we showed how exotic it could become (Figure 22).

Overall, our excitement was unbounded. Every working scientist, scholar, or detective of anything, for that matter, knows the rush that occurs with discovery. Another secret of the natural world revealed, and you were there, front-row seats and all. It was an exciting time, and more changes were on the horizon. I had decided to take a job at Cornell University Medical College and was moving back to New York City.

CHAPTER 5

BRAIN IMAGING CONFIRMS SPLIT-BRAIN SURGERIES

The scientist, by the very nature of his commitment, creates more and more questions, never fewer. Indeed the measure of our intellectual maturity, one philosopher suggests, is our capacity to feel less and less satisfied with our answers to better problems.

—G. W. ALLPORT

I JOINED THE FACULTY OF CORNELL UNIVERSITY MEDICAL COLlege at a time when doctors didn't advertise their services and money wasn't the 24/7 topic of hospital employees. Medical schools were exciting places to be and the doctors worked hundred-hour weeks without blinking an eye. It was the tempo and the intensity of a first-rate medical school that captured my energies. I loved it and I knew I was going to learn a lot.

The first big thing was that I had traded graduate students for medical residents, who were completely different animals. Graduate students are trained in the experimental methods of science, on how to do experiments. Residents are a bit older and wiser. They are making more decisions in a day than most of us make in a year. They interact with people dying, rejoicing, crying, laughing, the whole gamut of life's emotions. In a word, they are seasoned in a way graduate students are not. My job was to help bring these two kinds of experiences and skills together to study human cognition. I was to now mentor both Ph.D.s and M.D.s

Fred Plum, the legendary chairman of the Department of Neurology at Cornell, was the catalyst. Somehow he had gotten it in his head that his residents needed to be trained in neuropsychology, and somehow he got hold of me when I was at Stony Brook. The first idea was to come into town on Thursdays and do special neuropsychology rounds with the residents. That was a bold idea since I didn't know much about the vast variety of neurologic syndromes. I had read about them all and had experience with aphasics, but actually examining all kinds of patients? How was I going to be professorial about that, and on rounds no less?

Working the rounds at Cornell quickly became one of the great experiences of my life. Plum's residents, all of them, were outstanding and some of the kindest and most fun-loving people I have ever known. They quickly figured out I was the rookie at rounds. In a way, they gracefully became the teachers and I became the student. I found myself loving the neurology wards.

Soon enough, I began to get the hang of it and started to suggest experiments that might reveal something new about a classic syndrome. Busy residents don't mess around. If an idea crystallized, they wanted to do the experiment right then. "Here," they would say, "let's take the patient down the hall to the storage room. We can set up a projector on the table in there." Or, "There is a patient on Six East with a global amnesia. Get the portable EEG machine. We can document her seizures and then give IV Valium to bring her around." They did all of this on top of their regular grueling workload.

It wasn't long before Plum decided his plan for adding neuropsychology to his neurology program was working. He offered me a full-time job as a professor. I loved the idea, and it came at a moment when my personal life was changing as well. The offer came at a time when Linda and I had decided to go different ways. She would stay in Stony Brook with our four daughters—a huge source of joy in my life—during the week, and I would come out and spend the weekends with them. That was very tough stuff, but to everyone's credit, it worked out for everyone involved.

LEARNING FROM PATIENTS
AND ACCESSING THE UNCONSCIOUS

I convinced LeDoux to join me in my new lab at Cornell, and together we tried to figure out what our next projects might be. One of these grew out of rounds. One of the residents was Bruce Volpe, a superb physcian and a human with a preternatural energy. He started to show us patients with lesions in their right parietal cortex with what seemed to me the most bizarre syndrome. You'd ask such a patient to look right at your nose. Then you'd raise your left hand, showing either one or two fingers, and ask the patient what they saw. They'd easily give the right answer. Do the same thing with the right hand. Again the patient gave the right answer. Now came the critical observation. Raise both hands, such that the left was showing one or two fingers and the right was as well.

A truly remarkable thing happened. These patients all denied the information being provided by your right hand. It was as if your right hand no longer existed. The phenomenon is called "double simultaneous extinction." It goes on all the time in the neurologic clinic and is a disorder of attention. After one gets over the astonishing reality of the syndrome, the question arises: What the heck happens to that information about the right hand that you know got into the brain? After all, when only the examiner's single, right hand was elevated, the patient easily named the number of fingers that were raised. When the information was suppressed by having both hands raised, was that information no longer accessible to the patient's conscious cognitive system? Or was it accessible, but the patient unable either to talk about it or to be aware that the information was being used to help him or her make a decision? Maybe this anomaly would provide an avenue into the unconscious. Volpe and LeDoux went to work.

While Volpe rounded up a group of patients with similar lesions who manifested this phenomenon, LeDoux designed the experiment and helped Bruce learn some of the tricks of the psychological trade. The critical experiment was simple. We planned to flash pictures into each visual field simultaneously and ask the patients if the pictures were the "same" or "different." Thus the patient was simply required to make one spoken response. Yet, to get it right, information from both visual fields would somehow have to be combined in the brain and, following that process, go to the speech centers for a response. The first step was to see if the patients could do the task successfully. The answer was clear. The patients, who would deny the presentation of information in their left visual field, nonetheless could use it to make the correct "same" or "different" judgment. As you can imagine, when the patients were asked what had been presented on the "same" trials, they simply said two apples or whatever the stimuli had been. When asked what was presented on the "different" trials, however, they could never name the picture shown to the "extinguished" field.[1] Bingo, the experiment worked. This experiment turned out to open up a small cottage industry of research. Put simply, we had demonstrated that information that could not be consciously accessed could nonetheless influence how a seemingly conscious decision was made. We were able to peek into the vast unconscious, the networks that most likely govern most of what we do. We were terribly proud of ourselves, and soon enough, others took up the idea and extended it in many clever ways.

THE JOYS OF MENTORING AND FRIENDSHIP

I don't believe in "training" graduate students. I believe in exposing them to possibilities, under the assumption that if they need to know something in greater detail, they will learn it on their own. That is how I learned whatever it is that I know. When people talk about training, they generally mean taking an amorphous mind and shaping it into something. It is the sort of thing that goes on at universities that are not yet in possession of high-quality students. It is *not* the sort of thing that should go on at serious centers of discovery. Mentoring, on the other hand, is productive, necessary, and enjoyable.

Mentoring now takes place years beyond graduate school. The intricacies of modern knowledge are so vast that the graduate school experience has become only a small part of the total development of a young scientist. Over the years, especially when I was at Cornell, my experience in mentoring was mostly with students at the postdoctoral level. Students generally came to me with backgrounds in psychophysics* or cognitive psychology,† wanting to study patients with neurologic problems. Patients with focal lesions usually caused by strokes (or what was called "broken brains") provide an excellent means to study the workings of the mind.

One afternoon, Leon Festinger and I were lunching at Dardanelles in the Village. At the time, his interests were turning away from psychophysics and moving into archaeology and human origins. He asked me if I might be interested in taking on one of his students, Jeffrey Holtzman. To sweeten the deal, he said he would throw in his expensive computer-based eye-tracking device (Figure 23). The device was way too useful (and expensive) to end up in storage somewhere. The eye tracker allowed the experimenter to present visual

---

* Psychophysics studies quantitatively the relationship between physical stimuli and the sensations and perceptions that they incite.
† Cognitive psychology is the study of mental processes such as attention, language, memory and learning, problem solving, and so on.

information to a subject and very precisely stabilize where it fell on the retina. This means, for example, that if a stimulus was presented in the left visual field, something we did daily with our split-brain patients, the eye tracker would track the eye if it moved and automatically reposition the stimulus being presented. As Leon casually pointed out, the system was useless without Jeff to run it. Being a sucker for high-tech gizmos, I told him to send Jeff up for an interview.

FIGURE 23. Jeff Holtzman with his eye tracker.

One never really expects a new acquaintance to become a truly close friend. At the ripe old age of thirty-nine, I felt as though I had already met all of my lifelong friends, and that everyone I was to meet from that point forward would fit into the second-rung category of "acquaintances." As if to illuminate the foolishness of that point of view, in walked Jeff, and, within a week, we were inseparable. I relished that friendship for six years, and then Jeff died of an awful disease. His death was like suffering a brain lesion—part of me was gone and forever unrecoverable.

The year he graduated from the New School, Jeff married Ann Loeb, a dazzling young lawyer. Her first boss was Rudolph Giuliani. Ann became an authority on the First Amendment, and she read the copy for every issue of *Forbes* and the *Daily News* before it was put to bed. She also tried to read Jeff's scientific papers. It was Greek to her, but she did it anyway. Her sense of humor surpassed even his, which is saying something. It served her well in surviving the high-stakes world of New York City law . . . and being married to Jeff. They howled together, and just when he thought he had gotten her on something, she upped the ante.

Being with Jeff was like being on *Saturday Night Live* all day long. We would start to giggle at precisely the same moment when listening to a lecture, and we would have to avoid eye contact, looking straight ahead and concentrating fully to prevent a major disruption from occurring. Occasionally, I would giggle when he didn't, or vice versa. What's going on? I would think. Is Jeff even listening? Has he fallen asleep? Usually not, and wonderful arguments would follow over drinks at the Rockefeller University bar. We had them almost every night for six years. His sassy wit was nonstop and his impudence was an art form. One night as we left the bar, he said to me, "I think about Charlotte when I go to bed. Who do you think about?" Charlotte is my wife. How can you ever forget someone like that (Figure 24)? I tried to top him but I never quite made it.

Naturally, we talked a lot of science. Jeff was highly quantitative, and the numbers had to be really good before he would make any claims as to the results. He could think of the flaw in any experi-

FIGURE 24. Jeff Holtzman and I soak up life in Ravello, Italy. Having lunch overlooking the Amalfi Coast on a warm summer day with a dear friend is my idea of heaven.

ment, and he often successfully challenged longtime views of the lab. He would agonize over an upcoming lecture, or suffer over an upcoming grant renewal, and all the time he was the best the field had to offer. Mentoring him consisted of little more than putting him in a cab to go home after a good time.

Our work often involved the use of a tachistoscope, a device that presents visual information to one half brain or the other. In order for the tachistoscope to work, one has to be a good fixator, that is, be able to stare at a point on a screen with great care and intensity. Many people find this task difficult, and we worked hard to develop that ability among our patients. This was part of the reason that we traveled up to New England at least once a month for six years in a specially equipped van, loaded with this kind of equipment. The patients' families were wonderful to us; they always fed us lunch as we chatted. Investigating psychological processes in human subjects is a tricky and sensitive business. You are probing the innermost workings of someone's mind/brain. One must always strive to communicate the deep respect and gratitude one feels to the subjects and their families for participating.

On one unforgettable afternoon, we had driven to a patient's home in rural New Hampshire. We were gazing out the dining room window on our lunch break when I spotted a cow lying in the grass, staring down the hill, seemingly in a trance. I idly commented to the patient's father on the cow and its contented situation. Jeff was busy making himself a second sandwich and I assumed had likely lost touch the conversation. As I was wrapping up my cow conversation, I said, "Still, why is that cow so content to gaze down the hill all day long?" Jeff shot back, "Beats me, but he sounds like a good fixator. Why don't we go set up the old tachistoscope in front of him and see what's going on?"

As the last word fell from his lips, his face began to redden. He stared at his plate wishing he were anywhere else. Usually, such faux pas were my department, and Jeff made me pay dearly for each and every one, so I intended to relish this opportunity. Turning ever so slowly in his direction I said, "What's that, Jeff?" Gathering himself together he said, "I said I owe you one." Both the patient and his parents howled. A few years later they wept when they heard the awful news about Jeff's death.

Our New England trips were long, and they provided ample time to explore our views on just about everything. Jeff always talked about Ann. He was so proud of her. In short order she was arguing cases on behalf of the *Wall Street Journal,* the *Daily News, Forbes,* and any number of other impressive publications. He was apprised of all the legal details, and he took me through every one of them. I would challenge him, but he knew all the answers. If I hit on something that was privileged, he wouldn't give the answer because he said Ann would kill him. I'd ride him, but he never gave in. I'd get frustrated and say, "So how does it feel to have your wife make more money than you?" He would say, "Great, great. I love her. . . . I can't afford not to."

Jeff very much liked for things to be logical and orderly, although he didn't particularly relish orderliness in others because it frustrated his unbelievable ability to see relationships. He was an experimentalist. No one was better at that game, and it drove him to hilarity. One day, the results of a particular experiment were different every time he ran it. I said something like, This is good because maybe we're getting close to what is true. He yelled back at me, "True? Are you crazy? I don't care about it being true. I just want it to be consistent."

He was extraordinarily giving and yet, at the same time, infuriatingly his own man. He helped the entire lab on every experimental detail, and those who didn't take his advice should have. In his own work, he wanted, above all, not to make an error in logic on anything he reported. Could there be a loophole in his interpretation of the data? He would worry all night for weeks about a talk he had to give, afraid that someone would find a flaw in his reasoning. I would chide him with remarks like, "So you're wrong. Big deal. We are all wrong at some level. This problem is too big for our miserable human brains to solve. All we are striving for is to be more right than wrong. We don't have to be correct." His response: "Bullshit." I would tell him he was a compulsive jerk, and he would say I was a vague, undefined son of a bitch. We would go have a drink and decide we were both right.

When Charlotte and I married, Jeff was there. Our official ceremony was in Judge Rena Uviller's chambers in New York, followed by an all-afternoon lunch in a private dining room atop the World Trade Center. The morning ceremony had been attended only by Charlotte's sister and our good friend Nisson Schechter, who also happened to be the judge's cousin. At one point, Nisson told us how Rena called him one night with a question. She was deciding a case, and the plaintiff and defense lawyers were both Jewish and driving her crazy with details. So she asked Nisson, who knew all words Yiddish, for the Yiddish word for something like the big picture. The judge thought if she could find the Yiddish for the big picture, she could break through to these guys. Nisson said he didn't know the word but he would find out. He called seventeen rabbis. None of them knew. Finally, he called his old rabbi back in Detroit, who said, "Nisson, there is no Yiddish for the big picture. With Jews, it is all details, details, details."

It was a dazzling day, so simple and so meaningful. Jeff guided us through the whole emotional space, making sure we didn't get caught up. Rena Uviller had qualities of mind and heart that accented beautifully the fact that the most important event of our lives was transpiring in her book-laden chambers. At lunch, we all buzzed and laughed so hard at so many things that we were positively giddy (Figure 25). Around two thirty, the judge said she would have to excuse herself, as she had to return to court to sentence a man who two years earlier had murdered one of his children. Since then, he had been out on bail being a model citizen and holding down two jobs to support the rest of his family. What to do?

I will never forget that moment. In a matter of hours, Rena had

FIGURE 25. Charlotte and I were married by Judge Rena Uviller. We followed the ceremony with a private lunch at Windows on the World in the World Trade Center. (*Left to right:*) Jeff, Charlotte's sister Deezy Smylie, and my good friend Nisson Schechter, Rena's cousin, attended as well.

performed our marriage, participated in the revelry, and was off to deal with a further matter of great complexity and import. Jeff had set the tone of jocularity, but he had also projected the fact that he was always ready for questions about the mind and the heart. Somehow, ending our marriage lunch with a social conundrum was uplifting to us all. Rena would not have introduced us to that dimension if Jeff, the stranger in the room, had not instantly been able to communicate a deep sense of dignity even through his humor.

And then, with mind-numbing swiftness, Jeff's health failed. He had had a persistent cough for a few weeks, and when he started to cough up blood, he went to New York Hospital and was immediately admitted. His wife was about to have a baby, and for the preceding few months they had been under the stress of getting their apartment remodeled, living in drywall dust and all the rest. We had attributed his coughing to a million different causes. The culture showed it wasn't pneumonia, even though the lung films suggested it. Jeff knew he was in a bad way, and he called his family and his closest friend, T.L., to his bedside.

Three days into his hospitalization, Jeff's father, a physician, told me that he didn't expect him to make it. I was shocked and outraged that a young man in the best hospital in the world could be dying. They ordered a CT scan, thinking it might be lung cancer. It showed nothing, and he continued to go downhill. A lung specialist was brought in, and a quick pulmonary exam showed that Jeff's lungs were inflamed. A biopsy finally brought the diagnosis of Wegener's granulomatosis,* an autoimmune disorder. The prognosis was dim: Massive antibiotics and steroids were immediately thrown at it, but Jeff kept sinking. On the way to have the lung biopsy, T.L. reports that Jeff gave him a thumbs-up sign and said, "So lung." He tried to cheer us all up with stories and gags, which Ann and T.L. brought out to the waiting room.

At five in the morning, after his biopsy, I found him in the surgical intensive care unit. He was full of tubes, so he couldn't talk, but we carried out a conversation, with his part in writing. All he was concerned about was Ann—he felt horribly for her. I told him he was going to make it, but he ignored me and kept on probing for Ann's state of mind. I promised him that she was fine and that I would take care of her. He told me to take my planned day trip to the University of Georgia, and a nurse came in to shoo me away. We smiled our good-byes, and I never saw him alive again. He died the next morning, ten days after he got sick. He was buried three days later, and the following morning his wife gave birth to their beautiful daughter.

We all struggled to cope with the loss through the following

---

* Now known as granulomatosis with polyangitis, an inflammation of the small- and medium-size blood vessels that affects many organs.

days, weeks, months, and years. Charlotte and I had our first child a couple of months after Jeff's death. We spent as much time as we could with Ann and her baby. I took up cooking as a way to get focused on something new. We were all numb for a long time. Emotions are difficult things to understand. It is said by some that emotions are managed by old, subcortical parts of the brain, and as such they are inaccessible to conscious analysis. This may be true. It is also true that emoting does not obviate moods. My emotions won't leave me alone, and simply thinking about all these things privately doesn't help, so I write these stories. I have something happier, though bittersweet, to report as I type this. Two weeks ago, some twenty-eight years after Jeff's death, Charlotte and I watched a radiant bride walk down the aisle, Jeff's daughter. The best part was that she was witty and irreverent and cracked jokes the whole time, even when it was the groom's turn to talk. There is no doubt she has Jeff's spirit.

Jeff was smarter than most, he worked harder than most, and he was charming like few people in the world. With all of that, and with all his scientific competitiveness, he was remarkably free of ambition. We talked about it a lot, but I never understood it until his funeral gathering. Jeff's friends came to New York from everywhere. We drank until we were numb. We stared helplessly at his beautiful pregnant wife, his dazzling mother, his spunky sister, his stately father. We talked, cried, planned, drank, laughed, and finally broke down. The truth was that Jeff didn't need to be ambitious. What sustained him were his friends. He had collected in his short life the most astonishing group of friends I have ever come across. Whenever his phone rang, he knew it was most likely someone he felt for, felt good about. He always talked about his other friends, but most of us had never met. Only at his death did we discover each other, and the grace of that discovery was that, through his friends, it was clear Jeffrey David Holtzman would live on.

HAVE VAN, WILL TRAVEL

But I have gotten ahead of myself. While Jeff was working in the lab, LeDoux's footprint at Cornell was expanding. He had decided to return to the original questions about emotion that fascinated him, and to his work with animals. As he did with everything, he plunged into the study with gusto and brilliance. It was only a few years before he would be known as the world's expert on emotions and the brain. This meant learning a whole new suite of research tools and a new literature. No problem for him.

Before he switched to another field, however, LeDoux had contributed a key paragraph to a grant application I was writing. The split-brain team wanted a proper traveling van for testing. When we moved to New York City, we had ditched the old trailer and borrowed an old school bus going unused at a Cornell's affiliated hospital in White Plains. We built our modifications into the rigid seats, but after driving the big yellow bus up to icy Vermont, we were done with it. It is truly amazing that we let America's children ride around in those tin cans.

So, in our application to the National Science Foundation, we listed in the equipment section one GMC Eleganza motor home. They were about thirty-two thousand dollars. I was laughing to myself as I typed it into the formal grant budget pages. Such an item would definitely need a budget justification. In walks LeDoux. "Joseph, I could use a little help with stating why we need the Eleganza." Joseph said, No problem, and disappeared for about an hour. He came back with a full-page rationale of why it was central to the program and why Eleganza, in particular, fit the bill. We needed it not only for the living area, which we would modify for our testing lab, but also places to sleep and eat, saving money on travel expenses. Into the grant went the Eleganza and its justification, and off it went to the foundation.

About nine months later I got a call from the foundation program officer. "I have some good news and some bad news for you on the grant," the voice said. "We will not be able to fund the research assistant you requested and, for that matter, your own salary. It is tight times, as you know. But the committee did feel the Eleganza was a good idea, so they funded that in full. In fact, that is all they funded. To be candid, it sort of sounded like a *Travels with Charley* story. We like it."

Well, we partied that night, and the next day, Joseph found an Eleganza at a dealership in New Jersey and went over to pick it up. I had to be at some meeting or other but when he gallantly returned with the brand-new vehicle we were giddy with excitement until it dawned on us. We were now faced with the greatest issue for any New Yorker with a vehicle: Where the hell were we going to park a twenty-six-foot van? Frantic phone calls were made. Finally, someone came up with a slim parking spot by a building on narrow Sixty-Eighth Street, between York and First, the building right next to Joseph's labs! As this was all happening, the idea was born that perhaps he could live there during the week, as housing was tight. There was another problem. How does one get into the parking place from that narrow street? (Figure 26)

Science is truly a team effort. We needed a backup specialist,

FIGURE 26. The tight parking place for the Eleganza is in the upper right corner. A very distinguished neurophysiologist told me once that the most important course she ever took in her life was high school shop. Practical knowledge shouldn't be underestimated!

and, by luck, I had had a summer job somewhere along the trail that taught me how to do exactly that, swiftly and adeptly, I must say. The gates to the parking slot were open, and up the one-way street I drove with cars parked on both sides. I pulled just beyond the opening, with Joseph and Jeff sitting shotgun. Charlotte, with the no-nonsense assuredness of a vibrant blonde Texan, held back traffic. Tough New Yorkers froze in their tracks, and I slipped the van into reverse and in one turn was able to back it down the slim driveway with a clearance of only four inches. Parking the Eleganza turned out to be my job for almost ten years. Oftentimes crowds would gather just to watch, and more than once they placed bets. At one point, the National Science Foundation officer had heard so many tales about trips in the van that he called to ask if *he* could stay in it over a weekend he was planning in New York.

The parking spot was next to an old city public-health building that had been taken over by Cornell. Right up the street on First Avenue was the most sublime Italian restaurant in the city, Piccolo Mondo. It was where we always went for either lunch or dinner when we found ourselves hosting visitors to the medical center, and we were great favorites of the maître d'. One day I arrived with Sam Vaughan, the distinguished editor from Random House, who upon stepping into the restaurant asked the maître d', "Where is your men's room?" The maître d' calmly answered, "Mine is in Brooklyn." Sam smiled and turned to me and said, "Everybody in New York is an editor."

On one occasion, I had been seated for lunch at a famous corner booth, which, we were told, was where Vladimir Horowitz ate almost every night. Dutifully impressed, I decided I should somehow return the kindness and proceeded to tell the maître d' about the new carbonara recipe I had mastered from reading Marcella Hazan's new cookbook. As my description rolled out, I noticed the maître d' beginning to look ill. When I finished, he said, "We do not make carbonara here anymore, but for you I am preparing a dish of it right at the table so you can learn how to truly make it." He did, and for the past thirty-five years, Charlotte and I have been making it at least twice a month.

Being in New York was like that: rich, unusual life experiences at every turn. The morning may have been spent on the hospital wards, examining fascinating patients with mysterious syndromes. Any day on the wards one might find a patient with an attention disorder such as the "double simultaneous extinction," as described earlier, or a patient with a fascinating aphasia, or early dementia, or a more ephemeral disorder like a transient ischemic attack, which meant you had to think fast to glean and verify the phenomenon you were studying before it disappeared. Even if you walked into the sunroom at the end of the hall, where patients sat to warm up and relax away from their hospital room, a surprise might occur. One day I introduced myself to a gentleman who, in turn, introduced himself. He was Paul Weiss, the famous Rockefeller University professor and mentor to Roger Sperry. I told him I had been Sperry's student, to which he warmly announced that Roger had been by far his best student.

One could get called away anytime for another opportunity to examine and evaluate interesting patients. Our success at Cornell was hugely dependent on the residents. We became a resource for them and they for us. As doctors roamed the hospital doing their chores, patient after patient was directed toward us. A beeper would go off, and there would be news from Payne Whitney, Cornell's psychiatric hospital, adjacent to New York Hospital, about a relatively young patient with Korsakoff's syndrome. This syndrome manifests with memory loss, confabulation, and apathy, a result of thiamine deficiency, and it is usually seen in malnutrition resulting from chronic alcoholism or weight disorders. Volpe would grab me, and over we would go to witness the confused man who had no idea where he was but was about to be utterly repaired by an IV injection of thiamine right in front of our eyes. Minutes later, a call back to the main wards: A woman with acute cognitive dimming was in need of assessment. From a scientific point of view, the neurologic wards are the most fascinating place on earth.

FROM SLEEPING RABBITS TO REAL PEOPLE

One of the most gripping procedures to watch was that of radiologists trying to determine which hemisphere in a patient is responsible for language and speech. Before neurosurgeons would operate in the regions near the language areas, they wanted to locate the language areas. Hemispheric variation was always a possibility, and, properly, they wanted to be sure. The radiologic procedure they used, done for purely medical reasons, was an opportunity for the neuropsychologists to learn a thing or two about the dynamics of interhemispheric processes. The procedure required the radiologist to thread a catheter through the femoral artery in the leg, up through the heart all the way to the neck, and the internal carotid artery, which feeds the brain. They would then inject sodium amytal, an anesthetic that put half of the patient's brain to sleep for approximately two minutes. After that, the doctors would withdraw the catheter a bit and rethread it up to the opposite carotid artery to test the other half brain. All of this was done under direct observation of the patient and the radiologist using a fluoroscope. Watching half of a human brain go to sleep is the eeriest experience I've had. It certainly trumped my earlier rabbit work.

What makes it a draining experience as well is seeing that a person's conscious state can be directly manipulated in such a dramatic way and always at some risk. In general terms, the patient is usually asked to hold both hands high in the air. As the anesthetic takes hold in one hemisphere, the contralateral hand falls limp. In the hemisphere responsible for language and speech, those functions are severely disrupted, yielding either total silence or gibberish. This is all especially dramatic because one knows the other half brain is awake, watching.

We were trying to answer a fairly exotic question. When the right hemisphere was home alone, so to speak, with the dominant left hemisphere asleep, could we teach it anything? Further, could it then communicate its knowledge to the left hemisphere after it had been awakened from the anesthesia? If memories were established in the right hemisphere when the dominant language system was asleep, could the left hemisphere language system, after it awoke, have access to the information that had been encoded while it was snoozing? In our experiment, we discovered the answer: no. At the same time, if the patient was asked to simply point to an answer on a card I held up, the right hemisphere (presumably) seemed to do just fine at remembering the encoded information. The information was in there, but it was stored out of reach of the language system in the opposite hemisphere.

NEW TECHNOLOGIES: CAN THE BLIND SEE?

It was such vibrant, fulfilling work. Yet nothing could match what we were doing across the street in our labs, where hard-core experimental science was pounding forward. Jeff had established Festinger's eye tracker, enabling unique split-brain experiments. In our prior studies, as I mentioned, we had sent information to one hemisphere or another by simply asking the patient to fixate a point on a screen and then quickly flashing the information either to the left or right of fixation. It had to be quickly flashed, because if it was left up on the screen for more than 150 milliseconds, the patient could move his or her eyes, thereby allowing each hemisphere to see what was being projected. The eye tracker changed all of that, ensuring that the image always remained in contact with the intended hemisphere. This meant we could show visual stimuli for longer periods of time. We could even show movies to the silent right hemisphere. Would the content of the movies affect the talking left hemisphere?

Soon two spectacular new patients arrived to capitalize on our technological advancement. Case J.W. was part of the Dartmouth series. His callosum had been sectioned in two stages, and he would prove to be extraordinarily interesting in every scientific and personal way. In addition, Case V.P. came to us from Ohio. She was part of another surgical series, headed by Dr. Mark Rayport, and she became exceptionally interesting as well. Throughout the remaining pages of this book, these two cases will be prominent. Overall, between the wards at Cornell and our growing group of split-brain patients, every day's work was like fishing in a stocked pond. Every time the experimental hook went in, up came another insight. It's no wonder we worked all the time.

In our early days at Cornell, Jeff had found the tracker to be a powerful aid to our routine use of the tachistoscope, and he applied it to patients without split brains. He'd become interested in a phenomenon called "blindsight," cleverly named by the distinguished Oxford psychologist Larry Weiskrantz.[2] Just as the name implies, it is a syndrome in which people who have lesions in their primary visual cortex can respond to visual information, even though they deny its presence. This isn't like the "extinguished stimuli" that LeDoux, Volpe, and I explored when we first started at Cornell. Those patients could see information if nothing competed with it in the opposite visual field. With blindsight, however, the patient simply can not see the object but can nonetheless point to or pick it up or react to it in some way. The many visual scientists, led by Weiskrantz, studying this believed the remaining capacity was due to intact secondary visual pathways kicking in and picking up the slack somehow.

The patients who had been written up in the scientific literature had not had the advantage of being studied with a fancy eye tracker. Only the tracker could ensure that a stimulus was placed in the visual field where the experimenter hoped it was and remained fixed there over a period of time. In other words, without the eye tracker, there was room for error in interpreting why there was remaining visual function. Once a region of blindness had been identified as having been caused by a central brain lesion, it behooved the experimenter to make sure that all stimuli were presented within the blind region and that none fell into any intact parts of the visual field that remained. That could only be achieved with an eye tracker, which Jeff had. All he needed was a patient to study. Sure enough, it wasn't long until one showed up at Cornell.

Jeff first studied a thirty-four-year-old woman who had undergone surgery to clip an aneurysm in the right half of her brain. The aneurysm was in her right occipital lobe, so the surgery was expected to cause blindness in part of the patient's vision. Sure enough, after surgery, the patient had a dense left homonymous hemianopia—she couldn't see to the left of a point she was looking at. She was given an MRI, which revealed an occipital lesion that clearly spared both secondary visual regions and the superior colliculus, the main midbrain candidate for residual vision associated with blindsight. These intact areas should have been able to support many of the blindsight phenomena commonly reported.

But the patient had no blindsight. Jeff studied her for months and got nothing. He wrote up the work and published it in one of the finest scientific journals.[3] It met with deafening silence. Blindsight was too big an idea to be shot down by one experiment, even a great, beautifully executed experiment. Jeff said, "Great, Mike: I come to your lab to learn some new tricks, and you know what I discover? Blind people are blind. That kind of brilliance ought to get me a job at Harvard." In fact, the broader claims about the nature of blindsight remain a topic for debate. Jeff soon moved on to more alluring questions.

Cornell had became something of a magnet in those days. On many fronts the work was taking hold in the scientific literature, and New York, well, was New York. Who didn't want to be in New York?

We caught the eye, for example, of the spectacularly creative Stephen Kosslyn and his student Martha Farah at Harvard. They met Jeff, and all were off on a scientific hunt for the brain basis underlying visual imagery, the processes that allow us to imagine and visualize objects and events in our mind's eye. Kosslyn, still in his thirties, was the world's authority on this fascinating question. It was logical to want to know how mental imagery might be affected by split-brain surgery. Jeff was pressed into duty.

The story was complex and involved all kinds of discrete, detailed experiments. The studies came at a time when the notion of modularity was emerging as a conceptual framework for viewing cognitive mechanisms. With a modular framework, complex mental processes, such as visual imagery, could not be thought of as monolithic, involving just one part of the brain. Instead, complex cognitive skills were now seen to be the end result of several interacting modules, which produced what seemed to be a unitary cognitive event. It is easy to say this, and though it is hard to provide evidence for that kind of thinking, Steve, Jeff, and Martha did just that. They saw that split-brain patients handled imagery differently in each hemisphere, thereby suggesting that each hemisphere had different modules available to process the identical stimulus.[4] Believe me, this is all you want to know about it.

New York is a place that draws people into its magic. One day, a letter arrived from Toronto: A young Italian scientist from Bologna was wondering if we had room for her in our lab. We did and Elisabetta Làdavas, to whom no word short of *vibrant* does justice, moved south into our lab and our hearts. Like all the Italian scientists I have known, she has a work ethic that is dazzling, and a lust for life that leaves everyone around her breathless. Fascinated by the problem of visual attention (like every one else that I seemed to have surrounded myself with), Elisabetta had a unique approach. Everybody wanted to know how visual attention was distributed across a scene. So, for instance, if vision were viewed as a TV screen, was there more attention on the right side of the screen than on the left? Was there more attention on the top part of the screen than on the bottom? As Elisabetta worked on this question with teams of scientists, she always added her own twist. How is visual attention distributed if you look at a TV by bending down and looking through your legs at the screen? And left becomes right and vice versa? I'll never forget the look of astonishment on Jeff's face when she proposed this; months of experiments ensued. To this day she remains one of our closest friends and has become a distinguished scientist, successfully breaking through the rather male-dominated Italian academic culture.

GEORGE A. MILLER AND THE BIRTH OF COGNITIVE NEUROSCIENCE

New York offered so many things, not the least of which was the talent at Rockefeller University and, in particular, George Miller (Figure 27). I had just arrived at Cornell and was seeking companionship with someone well versed in psychology. Right next door was Miller, one of the few giants in the history of psychology, so I called to ask if I might come over sometime. He said sure, and suggested we have

FIGURE 27. George Miller visiting us at our weekend home in Shoreham, Long Island, New York.

lunch. I had no idea this would lead to our developing the field of cognitive neuroscience.

Both Miller and his office intimidated me. Not only did the office contain more books and journals than entire psychology departments, but it looked as if most of them had been read. As he stood up to greet me, I was surprised to see that he was as tall as me, which is to say, way over six feet. With little ado, we went upstairs to the Rockefeller Faculty Club—home to great minds and mediocre food. We took our trays of soup and sandwiches and sat down. As we tiptoed around various subjects, he occasionally inter-jected hospitable questions, like "Would you like a beer?" I said, "No thanks." Awhile later, he asked, "Would you like a cigarette?" I declined. A little later, he asked, "Would you like dessert?" Again, I declined. My thought was to keep things in the realm of pro-fessional simplicity. He looked at me in obvious exasperation and wondering, no doubt, if I indulged in anything at all, finally asking me, "Do you fuck?" I was silent for a moment, and then burst out laughing. Then I had dessert.

The ice had been officially broken, and I realized that George's reputation as a formidable mind had gotten the better of me. Char-acterizations of first-rate thinkers tend to take on a life of their own, with the result that neophytes like me begin to think these great personages would rather have a beer with an old friend than meet someone new and challenging. George put all that to rest with one hilarious crack, and within weeks we were good friends. Although I learned in the years to come that his reputation for unceremoni-ously dismissing faulty arguments was well deserved, I also learned that his comments, whether positive or negative, were inevitably con-structive with regard to good science.

Pierre S. DuPont made a wonderful observation to the French National Assembly some two hundred years ago. "It is necessary," he said, "to be gracious as to intentions; one should believe them good, and apparently they are; but we do not have to be gracious at all to inconsistent logic or to absurd reasoning. Bad logicians have committed more involuntary crimes than bad men have done intentionally."[5] That sentiment is the essence of George. He rarely talked about the personal dimensions of a given advocate, but simply observed if their reasoning was valid. When introduced to a body of information, his enormously quantitative and logical mind began a kind of digestion process, the outcome of which would be either encouraging or damning for the topic at hand. The product of this natural capacity is a rare scientist who could break from the conventional mode of thinking in a field and form a clear image of how things should be done. Again and again, George ventured into uncharted territory and produced classic papers that were harbingers of the vast activity that was to follow in this area of inquiry. Although his roots were in psychophysics, his main intellectual concern had always been the psychology of language.

In his earliest work, around 1950, he examined the perception of language, borrowing a host of technical tools from engineering. These included information theory,* which provided a rigor that had been previously unattainable in the psychological study of language.[6] In what was to become his signature style, he first drew a host of colleagues and students into the study of language perception. After establishing the importance of meaning and redundancy, he then followed that lead by shifting his interest and attention to language comprehension.

At about this time, Noam Chomsky released *Syntactic Structures*,[7] and George was quick to see its implications for the psychological modeling of comprehension. He immersed himself in Chomsky's writings. He and Chomsky spent six weeks together with their families in one house, during a summer course at Stan-

---

* Information theory is concerned with the quantity and quality of information and is a branch of applied mathematics, computer science, and electrical engineering. Formally introduced by Claude E. Shannon in 1948 with his classic paper, "A Mathematical Theory of Communication," it was developed to solve the problem of how to transmit information over a noisy channel.

ford University in 1957. George described, in a brief autobiography, what a daunting experience that was for him; given the caliber of George's own mind, that statement gives a clue as to how much of a genius Chomsky truly is. George's work during the next few years, exploring the relationship between transformational grammar[*] and comprehension, placed the field of psycholinguistics on a sound footing.[8]

George, who died in 2012 at age ninety-two, spent his life tugging back the curtain that obscures the secrets of language, and in doing so, he not only led the field of psycholinguistics, but restructured the field of psychology. Through the study of language, he learned and taught the rest of the psychological community that when describing behavior one cannot ignore the processing that mediates stimulus and response. Meaning, structure, strategic thinking, and reasoning are too large a part of even the simplest perception to be ignored. George and a few other seminal figures, such as Festinger, Premack, and Sperry, are responsible for changing the face of psychology: transforming it from a science of behavior to a science of mentation. Nevertheless, what has fascinated me over the years is that George, a highly rational person, did not approach his new endeavors with much forethought. Like most great scientists, he became interested in some phenomenon or other and then simply jumped in to try to illuminate the problem. As a story develops, either a new insight is gained, or the idea is a bust.

My own years with him were filled with another enterprise: launching yet another field, which has come to be known as *cognitive neuroscience,* the study of how the brain creates the mind.

---

[*] Transformational grammar is a theory developed by Chomsky of how grammatical knowledge is represented and processed in the brain. The idea is that each sentence in a language has both a deep and a surface structure. The deep structure represents the relations between the words of a sentence, and is mapped onto the surface structure via *transformations.* Chomsky believes there are considerable similarities between the deep structures of all languages that surface structures conceal.

It was born of rather intense interactions based primarily at the Rockefeller University bar. For about three years, George and I met there regularly after work and talked about our fields. He always had a deep interest in biology and assumed that much of psychology eventually would be an arm of neurobiology. A major problem with the then-current state of affairs was that neurobiologists, almost without exception, assumed that they could talk about cognitive matters with the same expertise with which they could talk about, for instance, cellular physiology. This is the equivalent of a textile expert talking as knowledgably about high fashion as she does about the pros and cons of polyester. It was unmitigated arrogance, and it drove many serious psychologists away from the brain and sciences, but not George.

We started exchanging stories, mine about episodes in the clinic and his about new experimental strategies. I would tell him about patients with high verbal IQs who lacked a grammar school child's ability to solve simple problems. He would tell me that psychologists do not yet have anything resembling a theory of intelligence or mind. He urged the continued collection of dissociations in cognition seen in the clinic, in the hope that a theory would emerge from these seemingly bizarre and scattered observations.

One day in the early 1980s, I took him on my rounds and showed him a variety of phenomena, ranging from perceptual disorders to language disorders. He had never seen anything like it and commented afterward that the neurologic patient was really what many psychologists were looking for. After all, he observed, psychologists try to test the brain's limits by making college sophomores work fast or by rapidly presenting them with stimuli to provoke errors. In the clinic, the errors pour out of damaged brain systems with little or no effort.

One patient we saw was a distinguished New York executive who had fallen down a staircase. He was reported to be globally aphasic, which means that he would not understand much, if anything, that was spoken to him and would speak only a little. As we arrived in his room, the computer tomography technicians were fetching him for a scan, so George and I tagged along. The technician asked Mr. C. to slide over to the gurney, to which he replied, "Yes, sir." Once positioned and rolling down the hallway to the scanner, he was asked about his comfort. "Are you feeling okay?" "Yes, sir," said Mr. C. After arriving at the scanner, the technician slid the patient off the gurney onto the table and again asked if he felt all right. "Yes, sir," said Mr. C. The scan was performed and Mr. C. was returned to his room. The technician, who was familiar with my studies, turned to me and asked why we were interested in this patient, as he felt there was nothing wrong with him. I turned to the patient and said, "Mr. C., are you the king of Siam?" "Yes, sir," he replied with great assuredness. George grinned and observed that success is always grounded in simply asking the right question.

It wasn't always fun, however. Once we started our formal program, a series of famous neuroscientists and cognitive scientists came for a week at a time to observe and share ideas. Accompanying this event was the obligatory social dinner, which included other scientists and neurologists. The intent at these dinners was to continue discussing the theme topics of the week, in a slightly less formal manner. Usually they were pleasant, even inspiring, but one dinner was an exception. There were about eight guests in a private room at the New York University Club. After a drink or two, we all sat down for dinner, and the soup had no sooner been served when one of the neurologists cleared his throat and said, "The history of the study of the human mind has been rich in neurology, but can you tell me one thing psychologists have discovered in the last hundred years?" I could not believe my ears. George solved the problem by thrusting his chair back and walking out of the room. What ensued was the longest and most awkward dinner of the decade. George and I never spoke of it, ever, but it served as an emblem of how difficult it would be to structure a new field.

As we continued to consider how best to launch our new field during our evening rendezvous at the Rockefeller bar, we talked about everything from neglect to neologisms. It was on one of those evenings, in a taxi leaving the bar, that we coined the term *cognitive neuroscience*. What we meant by cognitive neuroscience would emerge, slowly. We already knew that neuropsychology was not what we had in mind; tying specific cognitive capacities to brain lesions would not be our enterprise. The intellectual limitations of that idea seemed self-evident, especially with the advent of new brain imaging techniques. These techniques were revealing that lesions previously considered limited to only the primary tissue they damaged instead had more extensive damage to the surrounding area. This meant that it was less clear what areas of the brain were performing what functions.

One evening I asked George, "Just what is it cognitive science wants to know?" He looked at me, alerted for action, and then said, "Let me think about that." The following week, the guiding ideas behind cognitive neuroscience took form in a long memo from him, which I present in edited form in Appendix I.

Somehow our ideas came together and we cooked up a plan. George had been advising the Sloan Foundation on the general topic of cognitive science. The foundation had always strongly supported MIT. Accordingly, it was considering funding MIT, where cognitive science was taking shape as a focus on linguistics, almost exclusively. Presenting the diagram shown in Figure 28, George convinced the

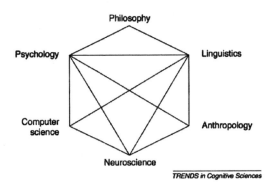

FIGURE 28. George Miller's long report on the state of cognitive science for the Sloan Foundation was summed up with this diagram. This simple summary of his hard work encouraged scientists to consider neuroscience as part of cognitive science.

foundation that the narrow linguistic view was shortsighted. He argued that the cognitive science(s) should be inclusive of related fields, one of which would become mine, "cognitive neuroscience." As he put it in a 2003 journal article:

> *The report was submitted, reviewed by another committee of experts, and accepted by the Sloan Foundation. The program that was initiated provided grants to several universities with the condition that the funds be used to promote communication between disciplines. One of the smaller grants went to Michael Gazzaniga, then at the Cornell Medical School, and enabled him to initiate what has since become cognitive neuroscience. As a consequence of the Sloan program, many scholars became familiar with and tolerant of work in other disciplines. For several years, interdisciplinary seminars, colloquia and symposia flourished.*[9]

They sure did. Jeff's wife, Ann, helped me set up a nonprofit 501(c)(3) called the "Cognitive Neuroscience Institute" and we convinced several New York universities to take part; a couple of years later, we benefited from an application to the Sloan Foundation for funds. Our goal was to facilitate cognitive neuroscience any way we could think of. We did it in several ways. We still do it.

SPECIAL MEETINGS, SPECIAL PLACES

One of my many personal paradoxes is that although I am about as routinized as one can be, I hate the status quo, especially the intellectual status quo. Helping to develop a new synthesis between fields especially appealed to me. In order to foster interdisciplinary interactions, once cognitive neuroscience studies were up and running I would hold an annual, weeklong, ten-person conference. Since I was a one-man show, my strategy was to pick a topic of great inter-

FIGURE 29. Time for a coffee break in Kusadasi, Turkey. Seated at the bar, from left to right, are Leon Festinger, Stan Schachter, Gary Lynch, Michael Posner, Steve Hillyard, and Ted Bullock.

est, pick a venue people loved to visit, and let each of them have a full half day to talk about their research. It worked. The venues included Barcelona, Kusadasi (in Turkey), Moorea, (one of the Tahitian islands), Venice, Paris, and Napa.

Of course, most of what happens at a meeting takes place in between the formal sessions. Each person is sizing up the field and asking questions that never get asked in the routine setting for professional gatherings. They are also, each in their own way, determining what is credible and what is not. It all comes pouring out during spontaneous lunches and dinners, walks through foreign villages, drinks at the local bars, tours of the local sites, and yes, occasionally, as a result of a question at the meeting itself.

One day, Festinger and his lifelong friend Stan Schachter, the social psychologist from Columbia; Gary Lynch, the enigmatic and driven molecular neurobiologist from the University of California, Irvine; and I were strolling through Kusadasi, a town on the Turkish Riviera noted for its colorful bazaars. We happened into a leather store that sold carry-on-size duffle bags. The bag had about twenty zippers on it. One could take a normal-size bag and reduce it to a handbag by continuing to zip it down in size, zipper by zipper. Stanley thought that was about the coolest thing he had ever seen and decided to buy one. Leon seemed impressed as well and was thinking about getting one, too. He was in the throes of a final decision when he suddenly said, "Wait, why and when would you use it?" Lynch fired back, "Oh, that's easy. Let's say you start out on a long trip and the bag is full of clothes. As you go along, you start to throw out your dirty clothes, and take the bag down a notch. By the end of the trip, it is only pocket size so you bring it home." It was one of those bonding moments (Figure 29) that are hard to get at the American Psychological Association meeting in Washington, D.C., with its more than eleven thousand attendees.

Lynch was a marvelous combination of raw intellect, endless curiosity, and just plain fun. He was a regular at our first meetings, as he had the essential ingredient: He could cut across local jargon and get to the ideas at hand. And he was witty. On the way to Kusadasi, I was changing planes at London's Heathrow airport and getting on a Turkish Airline plane to Izmir. There was Gary at the same gate, having just arrived from Los Angeles. As we settled into our seats in a row overlooking the wing, Gary turned to me and said, "You know that part of the wing that says 'no step'? All I see are footprints around it." We were truly off on a new adventure.

We sponsored a series of unforgettable meetings, each organized around a forward-looking scientific topic, such as the neurobiology of memory. The meeting on that topic particularly stood out. We held it in Moorea, as I'd spotted a fantastic travel deal: round-trip from Los Angeles to Moorea, with hotel, $770. It was in Tahiti, and the hotel was exquisite, nestled by the sea with sumptuous-looking food. So I cooked up another dream list of participants and got on the phone. "Hi, this is Mike Gazzaniga. We are holding a week long meeting in Moorea. We can contribute a thousand dollars to the costs. Would you like to come?" Ten invitations, ten immediate "yes" responses, all in about ten minutes. A few months later, Francis Crick; Geoffrey

Hinton, known as the "godfather of neural networks"; Corey Goodman, molecular neurogeneticist; Gary Lynch; David Olton, an expert on memory; R. Duncan Luce, the mathematical psychologist; Herb Killackey, an expert on neurodevelopment; Ira Black, a neurologist and basic scientist; Gordon Shepherd, an expert on neural circuitry; and of course my sidekick Leon were hot at it under the swaying coconut palms (Figure 30).

When Francis Crick was present anywhere, the chances were that the mean IQ of the room jumped up. His sparkling blue eyes and his incessant interest in biological mechanisms kept everyone alert. He was new to the field of neuroscience, which only meant he questioned more intently. After every speaker spoke, Crick would return to what became his mantra for the meeting: "But what you are doing is solvable in principle. The question is, what does it mean?" Let me tell you, that is an annoying question. Everyone was mumbling about it back in their grass bedrooms. What do you mean "it's solvable in principle"? Neuroscience was still trying to get the underlying data about basic brain functions. It was collecting the facts upon which a grand theory of what it all means would be

FIGURE 30. Moorea, Tahiti, the site of one of my small meetings with big minds. Francis Crick and I worked here, along with (*right to left*) Leon Festinger, Herb Killackey, R. Duncan Luce, and Ira Black.

built. Francis Crick and James Watson had already solved what a bunch of molecular facts meant for mechanisms of heredity.[10] Neuroscience simply was not there yet. Today, some thirty years later, neuroscience still has not collected the key data because, to some extent, it is not known what that key data even is. By the end of the meeting, everyone present had a far deeper sense of the issues and appreciated the conflicting views.

Years have intervened, but the idea that neuroscience needs cognitive science has prevailed. The molecular approach, in the absence of the cognitive context—which is to say, studying the brain without the mind—limits the industrious neuroscientist to pursuing answers to biological questions in a manner not unlike that of the kidney physiologist. Although such approaches represent an admirable enterprise, when put in that light they make it impossible for the neuroscientist to attack the central integrative questions of mind-brain research. Cognitive neuroscience has now become something of a household word, with its own journal, society, and conferences. Some of the most highly attended meetings at the huge Society for Neuroscience convention are on topics in cognitive neuroscience.

THE TWO POSNERS, ONE OF A KIND

I never met Michael and Jerry Posner's parents, but they did something right. These two brothers, Mike being one of the world's leading basic scientists of the brain, and Jerry, one of the world's leading neurologists, are spectacular intellects and, even better, spectacular human beings. Mike's home base is the University of Oregon in Eugene, a town he dearly loves. Yet he was eager to travel if the visit would help satisfy his unending desire to do research on how we humans work. A student of the great Paul Fitts at the University of Michigan, Mike was now his own man and had decided that maybe the new field of cognitive neuroscience might be of interest. George Miller and I had launched a very young program on the topic, and

Mike came to New York to help. Having his brilliance around made a huge difference. It also didn't hurt to have his famous brother across the street at Memorial Sloan Kettering keeping an eye on us.

Mike's passion was the phenomenon of attention. How did it work? Would it be disrupted in patients with various kinds of brain lesions? Indeed, what happened to cognitive processes following brain injury? In the early 1980s, brain imaging was not yet available for clinical use. Brain-lesioned patients were all we had. So, like George walking the wards, Mike thought that hearing a series of experts talk about all of this would be elevating. Of course, back then, sitting next to him in various seminars, grand rounds, and more was Jeff. The precise and well-formulated paradigms Mike had developed appealed to Jeff, so Jeff started to parrot me, asking Mike: "Why don't we try your stuff on the splits?" This became the cry, not only for Mike's work but for many other scientists who passed through our seminars in those early years.

Posner had basically shown that processes of attention could be defined and quantified precisely. For example, a person could direct their attention to a spatial location and if they did, then a subsequent event at that location could be responded to more quickly. If the person had been tricked, and the experimenter had told them to attend to a particular place, but the subsequent event, known as a *probe,* came up at another place, the reaction time was much slower. Attention was indeed a process that manifested itself clearly in Posner's elegant paradigm, and understanding how it worked would be cool. Jeff asked, "What if we cue the right hemisphere to attend to a spot that only the left hemisphere saw. Would the right hemisphere then also be better at responding to a probe?" In other words, could one half brain get the opposite half brain ready for an event, even though the other half brain didn't know it was being set up? In short, is the attention system somehow still connected in a split-brain person?

That is exactly what Jeff found.[11] Somehow a disconnected half brain could alert the opposite half brain to get ready for something to happen. It could not tell it what to get ready for in a cognitive/perceptual way. It could only tell it to get ready. It was a solid finding, and one that Mike Posner found intriguing. Apparently studying complex mental skills, such as attention, could be illuminated by studying patients, in this case, split-brain patients. He returned to Oregon and soon established a pioneering relationship with Oscar Marín, an outstanding behavioral neurologist in Portland. For years Mike went to Portland weekly to study patients. He was indeed hooked. A few years later, he went to Washington University to help launch some of the first studies on cognitive processes that used the new brain imaging techniques being developed by Marcus Raichle and many others.[12] It was becoming hard to keep up in our expanding field.

As other luminaries passed through New York, it fell to me to entertain them for dinner after their day's work. The usual drill was that George, Charlotte and I, the guest, and two or three residents or postdoctoral fellows would go to dinner at a First Avenue restaurant, such as Piccolo Mondo or Maxwell's Plum, or the Manhattan Club, or even on occasion, Mortimer's. Now, a dinner in New York, even in those days, could not be had for the twenty-five dollars a head we were allotted. We were always cognizant of costs, but we represented a major institution and foundation, and we were dead set on hosting in the proper fashion. The reality was that the dinner bill ran around sixty dollars per person, which we submitted for reimbursement. This went on for a year or so.

One day, I got a call from Plum's assistant, Gertrude. She informed me that Dr. Plum had decided that there would be a twenty-five-dollar limit per person for dinners involving the neurology department and that was that. I complained to Jeff about the new unrealistic rule, to which Jeff shot back, "Oh great. So when we go out to dinner, we say, 'Excuse me, Dr. Kandel, would you please skip ordering an entrée?'" After puzzling the problem, I instructed my secretary to write a memo to the file stating that Dr. and Mrs. Konstant would always be joining us for our hosting dinners. Our ghostly guests would help cover some of the shortfall because the restaurants were not going to give us a break, needless to say.

Years went by before I got another call from Gertrude. Our visitor's program had long since ceased, and we were on to other adventures. Apparently, Dr. Plum himself had finally taken his own group out to dinner, received the New York–size bill, submitted it for reimbursement, and was rejected. He wanted to know, "How did Mike get reimbursed all those years?" I pointed out to Gertrude that Cornell did not have a limit for such things, that it was Dr. Plum's own limit he asked accounting to impose on the neurology department. All he had to do as chair was to call up and lift the restriction. "Or," I said, "put down that Dr. and Mrs. Konstant were there as well."

BRAIN IMAGING CONFIRMS SPLIT-BRAIN SURGERY

A major question persisted in our now twenty-year-old work on split-brain patients. Were they actually completely split? Did the surgeons cut everything they said they cut? Was the entire callosum sectioned or had there been some inadvertent sparing of connecting fibers? A surgeon's notes of an operation and the actual reality of what was done inside the head can vary and frequently do. Thanks to computer-driven microscopes and more, this problem has been effectively addressed over the years and is a story in itself. Still, for us the question was simple enough. Are the patients completely split?

As if there weren't already enough excitement in our New York lives, the field of medical brain imaging was emerging with such lightning speed and with such proximity to us, we were about to find out the answer. Of course, CT scans had been around for a few years. Their ability to detect tumors and other abnormalities in the head and body were already legendary. At the time, however, they could not detect the white matter—that is, the communicating nerve fibers—of the callosum and couldn't help with our question.

Yet, right on CT's heels came the development of magnetic resonance imaging (MRI), the imaging technique that was going to transform medicine and, to some extent, the whole field of brain science. Once again, I knew nothing about that form of brain imaging, and, once again, the clinicians at New York Hospital were our enthusiastic teachers. Professors Gordon Potts and Michael Deck took us under their wing and, before you knew it, we were scanning our patients and determining if they were truly split.

These were early days and, luckily, we were in the hands of pros. As fantastic as MRI was and would become, those on the forefront of the technique were feeling their way, trying to set the parameters of the scanner to obtain the best images for white matter. After much experimentation, Deck and Potts were ready. J.W., our star patient, slipped into the scanner. Would years of study have to be reinterpreted because J.W. was not fully split? Would he be as described by the neurosurgeon years before? The tension in the viewing room of the scanner was palpable, with Jeff adding that extra edge of angst.

J.W. was lying quietly. The machine was banging away, as MRI machines do. (Put in the most simplistic terms, the machine works by sending out radio pulses to enliven water molecules in the brain. These newly activated molecules quickly relax back to their normal state. The device's giant magnet picks up the change and the data is reconstructed to produce the brain image in the viewing suite.) Potts and Deck had chosen to view the first set of data in the sagittal view, a slice oriented from the nose to the back of the head. An image smack in the middle of the brain, exactly between the two hemispheres, should reveal a big black hole where the glistening, white callosum used to be.

The images started to roll into the viewing room. They started on the right side of the brain and slowly made their way to the center cut we were looking for. It is quite a sight, magic almost. Everything that came before that key moment had been figured out by hundreds of scientists from multiple fields, building bit by bit over the years.

To name a few, this included the help of exceptionally clever bioengineers and medics, who knew something about the body and the questions that needed to be asked, and computer scientists and physicists who figured out the calculations of unbelievable complexity that the computer was performing as we waited: cooperation on the grand scale. And, oh yes, living and working in a culture that encourages advances doesn't hurt, either. It is a beautiful thing, when you pause for a moment to think about it.

The images piled up one slice at a time through the brain, the recording parameters of a certain process called "inversion recovery" doing their job. The other nerve fiber tracts of the brain were visualized, the ones that should have been intact within the right hemisphere. As the images approached the midline, the white fibers were drawing themselves together to cross over the callosal bridge to the other hemisphere. We all held our breath: Would the fibers stop being visualized or would they continue across the bridge? Would parts still be crossing while others cut? Would there be splotches of white or would it be vast darkness? The image appeared.

It was black as an eight ball, all of it. The callosum had indeed been completely sectioned. Even better, the smaller anterior commissure, the connection left uncut in the Dartmouth series of patients, was sitting intact in the blackness like the North Star. Jeff and I just stared at each other. It was unbelievable. Our baby, cognitive neuroscience, had taken another small step forward. Converging evidence was going to make us all better at our science.

I can remember the radiologists asking if they had helped and if could they do more. There were great turf battles going on between radiology and neurology departments to decide who would manage the exploding scanner technologies. These discussions were based on the economics and tensions of medical school financing and how it is organized to get the bills paid: It has to look seamlessly organized to the public. Yet none of this affected us or the desire of the medical hierarchy to get exciting research done. Cornell was a world-class place and that was evident at every turn. So Jeff and I said yep, they had helped us quite a bit, and that we had two more patients we needed to study, P.S. and V.P.

P.S.'s images retold the J.W. story. He looked fully split on the MRI scan, except for a small "nodule" in the posterior callosum. After hours of exploration by Potts and Deck, recutting the MR data in several different ways, they concluded the nodule was an artifact of the machine itself and not neural tissue in any way. As with J.W., the anterior commissure, the smaller neural bridge between each hemisphere, was easily visible in P.S.

We also brought in Case V.P. from Ohio. She was becoming another star patient, and we needed to know her status. Years after she was studied at Cornell, she was also studied at Dartmouth. The Cornell studies showed that some fibers remained in what is called the genu of the callosum, the anterior—that is, forward—part that interconnects part of the frontal lobe. We also saw a suggestion of some spared fibers in the posterior callosum, a region we knew was rich with fibers that played a major role in the transfer of sensory information. The imaged connection was slightly forward of where it was known that visual information transferred between the hemispheres. We had not seen any evidence of transfer on our neuropsychological tests of V.P. for visual function, but we were concerned. From that day forward, we made a special effort to doubly examine and look for any transfer.

After we had moved to Dartmouth years later, and MRI research had advanced, we checked V.P.'s callosum again and did another scan. A new methodology and a new team of scientists had come along. The methodology was diffusion tensor imaging (DTI), where an MRI scan could more precisely detect the presence or absence of nerve fibers. The brain imaging team was headed by Scott Grafton, one of the most talented brain imaging experts in the country. Probing the callosum, especially at the key places where we thought we had found sparing of fibers, it became clear that there were indeed no fibers in the posterior region. At the same time, the anterior fibers were real and could be easily visualized and tracked. This meant we had an opportunity to try to figure out what a small, isolated part of the frontal lobe fibers might be communicating between the hemispheres. But that comes later.

Overall, human split-brain research, thanks to the new technological developments, was on much surer footing. Within a few years, the Caltech patients were also given MRIs, and the evidence was good that their callosums had been completely severed as well. There remains doubt, however, as to whether the anterior commissure had also been sectioned, as the imaging machine used was not the kind that always picked up its signal—according to the authors who published the findings. Still, the evidence was excellent for the field. For the most part, split-brain patients were indeed split.

WORKING AND PLAYING

Even with the fast-paced science, the development of a new field of science, and my first trade book,[13] my social life also managed to stay apace, especially with Bill Buckley. During frequent lunches with Bill and his friends at his favorite Italian restaurant, Paone's on Thirty-Fourth Street, plans were always being made. I told him he should have a screenwriter take a shot at one of his Blackford Oakes books and sell it to Hollywood. He said, great idea. I said I know a young playwright. He said he knew the agent Swifty Lazar. Quick as a flash Bill hired my friend, who was the husband of one of my neurology residents, and we were off to the races, sort of.

My biggest success with Bill was introducing him to word processing and, as soon as they came out, laptop computers. He would come over to my Cornell office and sit down at my Digital word processor, a huge clunker by today's standards, but a slick device in those days. He was amazed and, of course, wanted one for himself. That was soon followed by his fascination with a letter I wrote him from Ravello on my new Sonycorder, a small keyboard device equipped with a tiny tape cassette. It recorded your work, which was then made available to your secretary's playback device. It seemed like a perfect device for travelers. I used it to write my first book, *The Social Brain*. Bill wanted it immediately, but that quickly gave way to other devices, which came so quickly that he finally got his own guru of electronics to help him. He was like that, and he was also always thinking of the other guy. After one of his office visits, he diagnosed me as too sedentary for my own health. He bought me a membership to the athletic facility around the corner from my office—One on One. All I can remember about the experience is my personal trainer kept explaining to me why some of my muscles were really sore after a workout. "The problem is, you never use those muscles, so they ache after a good workout." If I never used them, I asked him, why was I trying to develop them?

But I wasn't completely sedentary. Charlotte's brother, Walter Dabney, was a park ranger at Mount Rainier and soon to be the chief park ranger of the United States. He was always urging us to take a climb with him up to the top of the mountain; I finally agreed because Bruce Volpe and the professor of neurosurgery, Dick Fraser, wanted to go. Charlotte wanted to go. Charlotte's sister wanted to go. Volpe's then wife, Nancy, wanted to go. Walter said he would take us up, but not until all of us could run four miles in thirty minutes. This was going to take some time. Every morning for a few months, we ran up the East River from our apartment on Sixty-Third Street, until we met the mark. We were ready.

The afternoon of our flight to Seattle, Fraser and I worked in a game of squash. He was an expert, I was a schlepper, but it was part of my new get-in-shape program. Toward the end of the match, one of my racquet swings went awry and hit Fraser over his left eyebrow. It started to bleed profusely. Fraser immediately said, "Don't worry. I will throw some stitches on it and meet you at the plane." After a couple of checkup phone calls, all was okay, and off we went, all of us arriving in Seattle a little after midnight and boozy. Were we really in any shape to scale the fourteen-thousand-foot mountain?

The morning after our late-night plane ride, we woke up to see the peak of the magnificent Mount Rainier out of our windows. The park housing was at five thousand feet, so the air was crisp and unusually warm on that spring day. Charlotte and her sister were up early, merrily packing all our stuff into the backpacks. The packs full, they tightened the cords on top. Proudly, they stepped back and decided to try one on for looks. This is when we realized that we had made a mistake common to greenhorn backpackers: They could barely lift the packs! It soon dawned on us that we had another problem: We hadn't run our eight-minute miles trussed with a sixty-two-pound backpack.

Somehow, we left about noon and somehow all of us made it up to Camp Muir, a mid-mountain station at about ten thousand feet. It had taken us more than seven hours, in large part because of me. Being almost six feet, six inches tall and weighing in at 230 pounds without the pack, I was asked to bring up the rear of the climb. Another facet of backpacking is that the tall guy with the big pack always ends up carrying other people's stuff, which adds to his weight. Walter and the others would form nice steps in the crunchy, early spring snow. The steps held for everyone but me. All were of normal size, but when I hit the steps, they broke through, causing me to have to lift my foot at least twelve inches high to get out of the hole and get ready for the next step. It just about killed me. I had to use some of those slacker muscles, and they let me know about it.

We didn't make it to the top. The team in front of us had had a bad accident. They were hanging on their safety lines down a crevasse, and two of their climbers had been killed. Walter was radioed that a helicopter was on its way to pick him up, and it was his job to get the bodies. "Why?" I asked. "Why risk your life for two people who are already dead?" "Because," Walter, explained, "those are somebody's sons, and they live in some congressional district, and that congressman votes for U.S. park funding every year. Anyway," he added, "it's the right thing to do." Minutes later, he ran over to sort of a private side of the mountain, took a leak, and came back to await the approaching helicopter. Over the roar of the whirling blades, I yelled, "How can you pee at a time like this?" "Because," he yelled back, "if the helicopter crashes, you want an empty bladder!" The people who make this country great are everywhere.

Who knows where ideas originate? We do at least know that the more diverse our experience, the more fluid our mind. Out of our Mount Rainier adventure, hiking, traveling, and socializing with a neurosurgeon, a neurologist, and a bunch of rangers, who are by their nature, "just do it" guys, came an idea that was related to my new fascination with computers, optical digital recorders, and training. Why not build a gadget that would allow young neurosurgeons to practice their skills on what would amount to a flight simulator for surgeons? There was a medically oriented foundation that looked favorably on innovative projects, so why not give it a shot? We did, and the money rolled in.

The task was to digitize a neurosurgical operation in real time complete with correct and incorrect moves, complaining anesthesiologists, mistaken adjustment, and even haptics—felt resistance to the control levers. The new technology permitted near-immediate access to all the video segments we had filmed, which were stored in digital files separate from the surgery video. This meant we could suddenly interrupt a seemingly normal operation with a visual crisis, like an unexpected bleed. We built it, we tested it, but the foundation, under the advice of outside consultants, withdrew their support. Of course, computer simulation is commonplace now. A passing interest and a passionate interest are two different things for sure. Still, the young program officer of that foundation, John Bruer, soon became the president of the James S. McDonnell Foundation and a huge supporter of our nascent field of cognitive neuroscience.

MOVING ON, AGAIN

A major priority was to spend every weekend in Shoreham, Long Island, where we had bought a fabulous and zany house that had been under continual reconstruction by its owner, Geysa Sarkany, a

Hungarian architect with an enormous and restless talent. After we bought the house, we continued to work on it until it was right for us. It became the center of our emotional life for years, a place where my four daughters could bring their friends, put on plays, hang out, and live.

Of course, living in New York City during the week, Long Island on the weekend, and needing to get to New England once a month led to many logistical challenges. It was becoming obvious another move for the family was imminent.

CHAPTER 6

STILL SPLIT

I learned very early the difference between knowing the name of something and knowing something. —RICHARD P. FEYNMAN

THE EARLY ENGINES OF COGNITIVE NEUROSCIENCE WERE now running strong, fueled by new methods of experimentation. MR images were everywhere: Amazing anatomical images gave researchers great confidence to claim whether a particular part of the brain was present or absent. Those of us who worked on white fiber tract systems, like the corpus callosum, were particularly exhilarated. Exact knowledge of whether nerve bundles were present (such as when fibers were inadvertently left intact) or absent allowed us to perform a fine-grain analysis to determine where in the nerve bundle information might be transferring between the cerebral hemispheres and just exactly what that information might be.

Added to this was the phenomenal development of the field of human electrophysiology (the study of electrical phenomena involved in mental processes), which scientists had turned from ho-hum into an exacting science. Leading the field was none other than my old buddy from Caltech, Steve Hillyard. After Caltech, he went off to Yale and trained with one of the senior psychologists in the field,

Robert Galambos. Together they were among the first to show how the simple, doctor's-office-ready EEG signal could be tailored to track the flow of information through the brain. They decided to start "averaging" the brain signals to see if a discrete brain response, when linked to perception and attention, could be detected.[1] Put simply, they flashed a picture to a subject and then made a short recording (a few hundreds of milliseconds) of the EEG signal that was produced. They flashed it again and recorded another EEG response. After doing this several times they added up all the EEG responses and averaged them. They looked to see if a discrete response could be detected that was linked in time to the picture that had been flashed. It could. These became known as *event-related potentials,* or ERPs; with these, a thousand ships were launched. So, now we had MR letting us know about place and ERPs letting us know about the temporal aspects—that is, the timing of brain activity.

A third churning in the field was the role of cognitive science. Sophisticated experimental psychologists were becoming more intrigued with the brain and, in particular, the human brain. Formally, the field of cognitive science advanced the idea that, contrary to what the behaviorists preached, we humans were not simply a big bag of stimulus/response associations. We had a "mental" life too. Not only was mental real, but it could be explored scientifically.

Lastly, the field of functional brain imaging exploded. It had begun with positron emission tomography (PET) and, a few years later, was enhanced with functional magnetic resonance imaging (fMRI). At the time, this development was limited to a few large medical centers, such as Washington University, Harvard, UCLA, and a few others, especially in London, but its seeds were everywhere. People were eager to see where it would go. In the early days of the studies, the work was centered on the blood flow changes within specific neural systems while simple perceptual or cognitive tasks were performed. Everyone was stunned that any of these things could actually be demonstrated. The entire field swooned.

So did my lab. We now had yet another powerful tool to study how the brain operated. It was going to allow us to make a more detailed analysis of the brain function of the patients with their callosum sectioned. Now we could study what one half brain was up to, without interference from the competing functions of the opposite half brain. In fact, the opposite half brain could serve as a control. We could try to find out whether perceptual or cognitive processes of one half brain were calling upon cortical and/or sub-cortical processes to carry out its activity. If fibers had been spared during split-brain surgery, we could study both their discrete functions and any information that they were carrying. And what would come about if we combined all of these techniques to study our patients? To say we were excited is an understatement. We were champing at the bit.

The small meetings I had organized were having an impact on my thinking and research. Not only did we do the regular meetings that I have already described, but we began to hold meetings in which we explored particular topics. These meetings focused on the early history of humans and what made them special, Leon's new passion. We explored areas rich in archaeological interest and pre-history, such as Jerusalem, Seville, and the South of France. Being part of those explorations was teaching me to paint a larger canvas when thinking about the brain. This kind of cross-disciplinary work is almost impossible to achieve within a university structure, where one is tied to a particular field of study both by inclination, time, and, yes, group affiliation, rivalries, and turf wars. Probing at the edges of scientific fields, or even trying to incorporate other disciplines into your own, takes major intellectual and social effort. What started out as a form of intellectual curiosity morphed into a serious agenda for scholarly growth. In the end, these special meetings proved to be a big part of my academic life. Their message was clear and motivating: Take risks at the edges of scholarship and seek integration. Most forays won't yield anything, but some certainly will.

We decided that we wanted to use all available approaches to brain imaging, and we wanted to do those studies with only the best scientists. In the past, we had welcomed them into our labs when they had simple computer-based tests to run. Now, for the heavier-duty brain imaging tests that required advanced, expensive equipment, we traveled to their labs with our patients. "Have patients, will travel" became our motto. That meant we could live anywhere. Did it really matter which airport you flew out of or into to get the work done? Forces were conspiring to move me from New York.

SIMPLIFYING OUR LIVES

Charlotte and I were the proud parents of two small children. Try as we might, New York City proved challenging, especially as we felt more comfortable living the more leisurely country life. We were also tired of driving the Eleganza up Interstate 95 once a month, year after year. While it had been tremendously fun for us and convenient for the patients, it was getting old. Our testing equipment was getting more complex, and the trips were not hospitable to our electronics. On almost every trip, we had to run into the local computer store for various repairs on the spot, as critical pieces had been jostled and broken by the journey. Meanwhile, the storage of the van back in New York was becoming an irritant. One of the postdocs was moving it and managed to back into a Rolls-Royce. I started talking up the idea that we get a place in Vermont, and Jeff got on board. Jeff and I were the ones doing most of the driving and testing and were convinced we had to do something. We could sell the van and cobble together a down payment for the small house, where we could both stay and test. This would cut down on our travel costs and benefit everyone. We egged each other on, even though deep down we still thought the idea had a snowball's chance in . . . ah, Palm Springs.

It was early in the spring of 1985 that we met our solution. On one of our trips, Charlotte and I saw the perfect house. Nestled in the woods, yet close to the main street in Norwich, it was just across the Connecticut River from Hanover, New Hampshire, where Dartmouth is located. We had looked at others in our spare time, but this one caught our fancy. Designed by a young South African architect and freshly built by two spectacular local craftsmen, Christopher Jackson and Michael Whitman, it was a post-and-beam clapboard cabinlike structure on ten acres. I fell in love with it even though there was one problem. The asking price was $195,000, a sum that seemed impossibly high for a research project. I took a picture of it and we returned to New York.

Phil Guica was Cornell's financial head at the time, and I gave him a call. He knew all about our troubles with the van. I told him all the logic of buying the house and the plan for covering the costs. He sort of listened but then simply asked, "How much?" I winced and said, "Well, one hundred and ninety-five thousand." He said, "One hundred and ninety-five thousand, for the whole thing? Do you know what a studio apartment in Manhattan costs? Sure, go ahead." And bingo, the deal was done. Then tragedy hit. Jeff came down with his lung disease in the next weeks and died. He never saw the place.

As we spent more time in Norwich, the Dartmouth community grew on me once again. It's the whole package, rooted in the sublime beauty of the Connecticut River's Upper Valley. Add to this the support of everyone from neurologists to psychiatrists. It wasn't long before my wife and I grew to love it.

My work at Dartmouth was also going well, and Alex Reeves, the chair of neurology at Dartmouth, very matter-of-factly asked me if I would like to move to Dartmouth Medical School. I blathered something. Before I knew it, Alex had arranged for an endowed chair and a position in the psychiatry department. Tenure at Dartmouth Medical School for a Ph.D. could only be granted to someone in a department that was part of the medical school. It could not be granted to a clinical grouping such as neurology. I had pretty much decided to take it when Fred Plum asked to see me.

Fred started in with a "life-is-bigger-than-all-of-us" voice and

I could see what was coming. So I said, "Fred, before you go on, I thought I should tell you I have decided to take a position up at Dartmouth. What with all the work I do up there, I thought it would make a lot of sense." Well, a big smile broke out, followed by a hardy round of congratulations, and all ended well. For all I know, Fred knew all about the offer and was going to nudge me to take it. After all, I already had tenure at Cornell and that was getting expensive for him.

Exhilaration always accompanies a move. People in different regions think about both academic and personal life differently. The fast, buzz-saw New York life gives way to the "aw-shucks, let's do this together" kind of life in a small town. Some of that is good and some of it can be a little jarring, especially if a move happens quickly.

SCIENCE IN THE WOODS

In the years following Jeff's death, and in the waning days of my life in New York, I turned into a bit of a loner and was somewhat depressed. We would go up to New England for long periods of time, and I started to do all the experimental preps and to run the experiments myself (Figure 31). It was a lot like being a graduate student again. At the same time, I was still running a large lab down at Cornell, and I had many responsibilities. Every lab of any size has key individuals playing almost selfless roles to keep it all working. Helping me keep these two lives together were all kinds of talented people, some of them with traditional training and some with unique styles and talents. Bob Fendrich was unique and everybody loved him. He represented the best in science and yet, paradoxically, was elusive, almost invisible.

Bob was another eye-tracker specialist and had also trained at the New School for Social Research. I hadn't realized that Jeff had been bringing Bob to our labs after work for the past several years. Bob would fix this or that and help Jeff in various ways. Right after Jeff

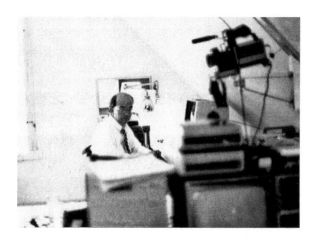

FIGURE 31. At my office in the Norwich, Vermont, house lab. In many ways, it was exhilarating to conduct all the experiments by myself: running the camera, programming the computers, designing the experiments, and administering the tests.

died, Bob and I met to discuss the possibility that he join the lab. I really knew nothing about him, so I gave him a grant to read and suggested that he get back to me with what he thought about the research. Within a couple of days, a five-page summary was in my hands, beautifully written, perceptive, and just plain first-rate. Bob, with all of his quirks, was now part of the lab. He brought with him tremendous scientific talent and standards. Bob was and remains the real deal and a wonderfully kind and gentle soul to boot.

With the formal move to Dartmouth, it was time to abandon my double life of two labs, two places to live, and the wear and tear that life inevitably brings. I needed a solid scientific group in the Upper Valley. I needed to recruit my team, and I needed to raise money to pay for it all. Life as a lone country scientist working away at his home-based lab wasn't going to cut it. And there was plenty to do. To my delight, Bob, though a veteran New Yorker, said he was in. Next was Mark Tramo, a young neurologist just finishing his residency with Plum. Mark was also a musician and had a driving passion to understand more about music and the brain. After a couple of delicious meals overlooking a dramatic waterfall at the Simon Pierce restaurant in Quechee, he was in as well. Kathy Baynes was easy to recruit. She was part of the granite of New Hampshire and had gone to Plymouth State College. Even though she had just received her Ph.D. in neurolinguistics from Cornell, she thought it was a great idea to head north. Another talented postdoctoral student, Patti Reuter-Lorenz, opted in as well. And finally, one of Hillyard's students, Ron Mangun, called. His wife wanted to do a residency in neurosurgery at Dartmouth. He asked, "If that works out, how about a job?" It did. A couple of years later, Ron became part of the backbone of my efforts. We started professional societies, wrote books, and got some great science done, too.

Of course, there is always the problem of space: None was available. We were all wondering what to do when someone figured out we could have the basement and first floor of an old white colonial house, Pike House, right across the street from Mary Hitchcock Hospital (Figure 32). The second floor was housing the AIDS program at Dartmouth, but the rest could be ours. We went to look it over. While we had our doubts that there was enough room, it looked sort of classy and was definitely one of a kind. We squeezed in. Fendrich took one of the basement rooms, and Tramo crammed a soundproof booth into the other. He didn't use it much, so Mangun took it over when he arrived for his electrophysiological studies.[2] Baynes took a first-floor room for an office, two more were set

FIGURE 32. Our beloved Pike House at Dartmouth.

aside for testing patients, one for a small seminar room, another for our secretary, and I had the back room. Tramo took a front room upstairs, and others were squeezed into corners. It worked.

All such enterprises, of course, require money. At Cornell we had successfully launched what is called a "program project" grant from the NIH. It was the first "cognitive neuroscience" program project in the country. It showed that the interdisciplinary approach to dealing with topics in mind/brain research could compete with the more traditional topics of straight neurophysiology or behavioral analysis. Such grants have five or six separately funded sections and, in the aggregate, amount to a lot of money. Fortunately, we were able to bring along the grant we'd used at Cornell, and, in due course, successfully renew it at Dartmouth.

As I already mentioned, we also brought along the Cognitive Neuroscience Institute, which, after all, was simply a checkbook that gave us access to funds for our unusual ideas. At one of these meetings, held in Venice, a golden moment occurred. We had met to think about evolution and the brain. Our group included the paleontologist and evolutionist Stephen Jay Gould; David Premack; Terry Sejnowski, who was the leading expert on computational theories of brain functions; Jon Kaas, the leading expert on comparative anatomy and brain evolution; Leo Chalupa, an expert on visual system development; and Gary Lynch, our expert on the cellular basis of memory and always a big thinker. Gould decided to make his presentation via walking tour through San Marco Cathedral. There we were, hearing from Gould his notions of adaptations and his spandrel theory beneath the very spandrels that had triggered his theory. (As Gould waxed on about the spandrels, people joined the group thinking they were getting a "free" architectural tour. They slowly began to look confused and drift away.)

A couple of years later we again gathered in Venice to hear more from Gould, and from Jean-Pierre Changeux, the leading French neuroscientist on brain evolution and function. Premack was there along with Steve Pinker, Wolf Singer, the leading German expert on visual function; Gary Lynch; Gilbert Harman, the distinguished philosopher from Princeton; the distinguished immunologist Manny Scharf; the psychologist David Rumelhart; and neurobiologist Ira Black. We'd met to discuss the challenge Danish Nobel laureate Niels Jerne had written about years earlier—the importance of selection versus instruction.[3] His idea was that maybe the brain, like the immune system, does not respond directly to the environment. Rather the environment, which impinges on any kind of biologic system, including the brain, selects preexisting (that is, innate) capacities. This was a radical idea.

It was an intellectual feast. Pinker stole the show with one of his first lectures on what would become his landmark book, *The Language Instinct.*[4] I remember Premack, dazed by how penetrating and lucid Pinker was, remarking, "If Pinker was any better, I would have just shot him." Indeed, years later Dave told me "You know, Pinker's book on language is the best book ever written explaining complex science." Dave is not freewheeling with his compliments. Everyone seemed to love those meetings. Overall, the idea of selection versus instruction is so powerful and the meeting was so provocative, that I spent a good amount of my time back in Hanover writing a book about it—*Nature's Mind.*[5]

LAUNCHING A SCIENTIFIC JOURNAL

Another idea was brewing. We wanted to start an academic journal, called the *Journal of Cognitive Neuroscience,* and we needed a publisher. I sent out queries to Johns Hopkins University Press, MIT Press, and the fledgling publishing house of Larry Erlbaum. I was hoping that one of the university presses would pick up the contract. I thought their prestige would encourage scientists to contribute. Scientists hate to think that their efforts may wind up off the grid in a sort of intellectual limbo, because their paper had been published in a journal that later closed up shop and was no longer accessible. While both Hopkins and MIT wanted to publish our journal, neither offered financial support. That was a problem: We had no extra funds. I was disappointed and complained to Leon. He told me to have lunch with Erlbaum, who was a straight shooter and a mutual good friend. Larry said let's have lunch at Piccolo Mondo.

We sat down to a lunch of calamari fritti—the restaurant fried each batch of calamari in fresh olive oil. (It always took twenty-five minutes to arrive at the table.) I explained my plight. Larry smiled and said that he would give us thirteen thousand dollars for editorial expenses. He told me to think about it and said he would send the contracts over the next day. I went back to my office and, after stewing awhile, let MIT Press know about the new offer. I remained concerned about the prestige factor. In seconds, MIT agreed to the same terms. I immediately called Larry to let him know, and he completely understood my decision to go with MIT. Larry went on to fame and fortune in the academic publishing business, and MIT Press has been an outstanding partner. It was my first scrape with a business decision, and while it worked out well for both parties, I didn't enjoy the conflict between cold economics and personal loyalties. Nevertheless, we had assembled a stellar group of editors and all seemed well.

Like everything else, the realities of the commitment were an entirely different matter. How were we going to start a journal that had not yet had any articles submitted? How was it going to be readied for printing? My wife and I looked at each other and said, "Looks like we do it." I got on the phone and twisted the arms of many a famous friend and asked them to submit an inaugural article. They all did, and the manuscripts started to arrive. Next step: How does this get printer-ready? Well, it was right about the time that the new program PageMaker had been introduced to the world, a program that let one electronically assemble, format, and deliver copy-ready material for printing. Charlotte became an expert. Final step, where

FIGURE 33.
Finishing the area
below the lab used
for patient trials.
This would become
the office for the
*Journal of Cognitive
Neuroscience*.
Jeff's close friend
came to visit after
his death, so we put
him to work.

were we going to do this? Once again we figured we could use a room in our Vermont cabin. We were running out of rooms, however, so we converted the garage (Figure 33).

We were beaming with our cleverness and decided to invite our associate founding editors, Ira Black from Cornell, Gordon Shepherd from Yale, and Steve Kosslyn from Harvard, up to Vermont. Ira and Gordon were molecular and cellular neuroscientists with a great interest in the nature of the mind. Kosslyn, perhaps one of the brightest and most clever cognitive psychologists of our time, held up that end of the equation. It was a cold winter night when they all arrived. Cars were slipping off roads all over the place. Yet nothing could daunt the energy of these determined scientists and their commitment to the project. Plans were made, standards were set, review requirements established, and workloads assigned. We had a big party, and we were all set.

Gary Lynch and Rick Granger had sent in a manuscript for the inaugural issue (Figure 34) that was full of math.[6] As Charlotte

FIGURE 34. The cover of the first issue of the *Journal of Cognitive Neuroscience* was a painting by the neuroscientist Alex Meredith, who coauthored a paper included in the issue. Each subsequent issue featured a fresh cover.

and I began to put it into the proper form for electronic delivery to MIT, we learned the hard way that PageMaker was not designed to do mathematical notations easily. It took us five days of fiddling around to get the manuscript into shape. Even though our manuscripts with simple text and figures were a snap, overall our first issue took way too much time. Something had to be done, so I decided to visit MIT Press and sort it out. MIT, I thought, one of the greatest scientific institutions on earth, would surely know how to make this new electronic printing easy.

It turned out we were ahead of the curve. MIT hadn't yet geared up for the new era of computer-based publishing. They still did it the old-fashioned way, using typesetters and human proofreaders. To our surprise, nobody in academic publishing had switched over. We managed to get two or three issues to them, leaving blank spaces in the text where the formulas were placed. To our glee, MIT simply told us to collect the manuscripts, and they would take it from there. They have been stellar ever since, and, of course, it's all done electronically now.

Successful journals take a lot of loving care to get going. Establishing a flow of high-quality articles is one trick. In the days of print, each issue was an event and each had to be balanced and attractive. We gave each issue a different, artistic cover, where we would trumpet artists and photographers. When putting together an issue, Charlotte and I would pour ourselves a cognac and climb into bed with our inbox of accepted articles. We picked the papers for the upcoming release that we thought made for interesting reading as a group, rather than in the order in which they were accepted for publication. It worked. To this day, the journal is a major success and one of the top-ranking journals in psychology and bioscience. This is in no small part due to the fact that Charlotte remains the managing editor.

MORE FUNDING, MORE RESEARCH, MORE KNOWLEDGE

Meanwhile, many visiting researchers were making a mess of the scientific literature on split-brain patients.[7] The researchers seemed not to understand how cross-cueing tricks, used by the split-brain patients to function in the world, could make it look as if their brains were not split. Many concluded, based on this misunderstanding, that there were noncallosal pathways deep in the brain exchanging information. The idea was that split-brain patients might not be so dramatically split after all. If these studies and conclusions were correct, the results of split-brain studies looked very different. Indeed, the very notion that the mind could be divided at its joints was up for grabs.

The proper testing of split-brain patients, while seemingly straightforward, is exhausting. Skilled experimenters have learned through experience and are well aware of all the strategies and tricks patients unconsciously and unintentionally use to cue themselves. They are cross-cueing all the time, and as a result, any experiment is at continual risk of not working properly. Even very sophisticated scientists can fall victim to appearances. We had several such episodes over the years. As a result, from time to time, erroneous messages were being reported and trumpeted in the scientific literature.

One very senior visitor to the lab was my old friend Donald MacKay. He visited while we were still in New York City. He and his physicist wife, Valerie, were dead set on showing that there were not two different agents in one disconnected brain. They had asked to test patients J.W. and V.P., and, of course, we obliged. They were both spectacular scientists and friends. They were not, however, attuned to the tricks of the split brain. The MacKays' task was to try to get one hemisphere to bet against the other. If it worked, they thought it would demonstrate the presence of two mental systems, each with its own "free will." They would accept the idea that there were indeed two mental systems, each with its own evaluative system, and that they were wrong. If it didn't work, they thought that would suggest there were not two minds, but merely two executive systems managing nonessential processing separately or something like that. They set up a clever test, but it was doomed to be nonconclusive because the split-brain patients could cue and control a situation in ways that undercut the experimental design.

In the MacKays' experiment the first order of business was to communicate to J.W. the idea of how to be a guesser. So in full view of both hemispheres, Valerie wrote a number from 0 to 9 on a piece of paper. J.W. could see it, but Donald, who was going to play the role of guesser, could not. The game would then be that Donald would guess a number, and J.W.'s left hand would be required to answer what he knew by pointing to one of three options printed on a card: "go up," "go down," or "OK." If J.W. had seen Valerie write, say, a "3," and Donald had guessed "7," the correct answer for the left hand would have been to point to the option "go down." This all proved very easy to learn and is a good example of what we had come to appreciate after years of split-brain testing. Always train the patient on the general strategy of the test before starting to examine how one or the other hemisphere might respond when asked to work alone. In short, the game of "twenty questions" was easy, and J.W. picked up this version of it very quickly. To wind up this phase of the test, Donald and J.W. switched tasks. J.W. was to be the guesser and to guess which number Donald had been privately shown by Valerie. Again J.W. learned to do this quickly and, of course, his responses were guesses since he had no way of knowing the number that Valerie had shown Donald.

Now it was time to test the right hemisphere alone. MacKay started off his experiment by asking J.W. to say what he saw when a number from 0 to 9 was flashed to the left visual field, and an alphabet letter was flashed to the right visual field. Of course, in standard split-brain style, J.W. could only name the alphabet letter that had been flashed to the left speaking hemisphere via the right visual field. Then MacKay asked the left hand to respond to his numerical guesses by pointing to the card with the three options on it "go up," "go down," or "OK." Again, J.W.'s right hemisphere picked it up fast. From the experimenter's point of view this is a tense time and becomes exhausting. After all, one is witnessing a nontalking half brain perceiving a number and choosing a response on its own terms and outside the realm of awareness of the chatty left hemisphere. One is seeing a module working independently, as if it were its own mind, yet still in the same skull as the left talking half brain. All of our intuitions are that this cannot be, and yet it is, right in front of our own eyes.

MacKay was now building up to his key test. Could things be set up such that J.W.'s left hemisphere became the "guesser," while the right was the keeper of the secret number? Here again, the secret number was projected to the silent right hemisphere. Instead of MacKay being the guesser, however, J.W. was told to let his mouth do the guessing, which meant, of course, his left hemisphere. Once again J.W. proved able to play this game of "twenty questions" with himself. This struck us as very eerie, before we realized that everyone probably does the same thing. Your conscious mind makes a guess, which your unconscious mind listens to. It then becomes active, which in turn makes you think of other associations, which finds you zeroing in on what an answer to a question might be.

The final step now arrived. MacKay set two boxes of tokens in front of J.W., one for the left hand and one for the right hand. MacKay then proposed to J.W. that the right brain (left hand) should be paid a token by the left brain (right hand) for every piece of information the left brain needed to solve the problem. Thus, after the secret number had been flashed to the right brain, the left brain needed some information as to whether a guess it (the left brain) had made was correct. Thus it would now cost the left brain a token every time it ventured a guess, to be paid to the right brain.

To keep things in balance, the left brain received added tokens, too, if it guessed the answer quickly. If the right brain made a mistake in its answers, it was penalized by the experimenter, who took three tokens out of its box (the left hand's box) and gave the tokens to the left brain (the right hand's box). That was to keep the right hemisphere honest and not mislead the left hemisphere, thereby requiring it to spend more tokens to eventually get the right answer! The game went on like this, and it all seemed to sort of work, but still nothing new was really being learned.

MacKay then threw some red meat into the game. Why, he asked J.W.'s right brain, don't you charge the left brain three tokens instead of one token for a piece of information? The right brain immediately agreed, which in due course depleted the left brain's token box, and the game was over. Why didn't the left brain (right hand) refuse the request for three tokens or embark on a bargaining session with the right brain (left hand)? Because nothing close to that happened, the MacKays felt they had preliminary evidence that each hemisphere did not have its own evaluative system and that, therefore, there was but one mind in the skull. They thought that somehow, information was being integrated across the midline through some unspecified subcortical system.

I disagreed. There were so many possible alternative explanations. The simple truth was that the reward values to the game, which they had assumed were being lateralized to one hemisphere, were likely shared by both hemispheres. It was already known that cross-cueing of emotional valence was a reality. The MacKays carefully noted my disagreement with their interpretation of the experiment when their subsequent article in *Nature* appeared a few months later.[8] Still, the paper attracted attention and others started examining other kinds of studies, which basically were testing for some kind of transfer between the hemispheres. The next visitor to the lab was a young neuropsychologist from Montreal's McGill University, the talented, dashing, yet enigmatic, and ultimately tragic Justine Sergent.

It is commonly observed between a dog and his master, between an old married couple, and between the horse Clever Hans and his trainer that a subtle cueing is going on all the time, most of it outside the realm of consciousness. It doesn't take too much for a trainer to cue his dog to stop at a particular point in space, which raises problems for validating the idea that bomb-seeking dogs are really doing the finding rather than following the master's hypothesis. In psychology, the well-known Clever Hans account, the horse that could do arithmetic, was all about the master unconsciously cueing the horse when to stop stomping his foot.

Couples and old friends can usually conclude a sentence begun by the other, having had so much practice together in social exchanges. In fact, couples can anticipate most everything, including each other's thoughts. With that in mind, how do you think two half brains coordinate to coexist together after months and years of minute-by-minute practice? My bet is they work it out quickly through having intimately shared expectations about the world they live in. While it may be difficult to predict the reactions of someone you just met, it is usually not too hard to predict those of spouses, children, fathers, mothers, and so on. It follows that it should not be too hard to predict behaviors of the half brain living next to your other half brain, watching the same world and feeling the same emotions, the same rewards and punishments of life. It is because of this that studying split-brain patients is difficult: What seems like central integration of information usually is not.

Still, the MacKays had floated an idea, and others thought that their ideas needed to be tested out. Having worked with these patients day in and day out for years, neither my lab nor I shared in that opinion. If correct, it would not only be news for split brain research but also surprise other researchers that higher-level processes could take place in subcortical structures. Sergent also wanted to pursue the idea that some kind of higher-order information exchange between the hemispheres was still intact following callosal section. She too did her work on one of our New York runs up to New England.

The long and short of her experiment was simple enough. First, see if each hemisphere could judge if consonants and vowels were different and keep a record of how fast they could do it. Easy, each hemisphere could, and vowels triggered a faster response than consonants. Then flash a consonant to one hemisphere and a vowel to the other, or a consonant to both hemispheres or a vowel to both hemispheres. In all cases, only one manual response from one hand was allowed—press one key if it was a vowel, and the other key if it was a consonant. Now, seemingly, with only one response allowed, the hemispheres had to exchange information, especially on the trials where one hemisphere was getting a consonant and the other a vowel. Since in this case each hemisphere had a different goal, how was the hand going to react? Because there was no incorrect answer on this test, the clues to underlying mechanisms could be gleaned only from the reaction times.

J.W. did not hesitate doing the task. If reaction times had not been measured, one would have simply observed that another visuo-motor task was being carried out easily. With the reaction times noted, however, another story was possible. J.W. responded most quickly when vowels were presented to both hemispheres. He was slower to respond when a vowel went to one hemisphere and a consonant to the other. Finally, he was slowest when each hemisphere was presented with a consonant. This pattern of findings led Sergent to the conclusion that there must be higher-order information crossing between the hemispheres. In her assessment, if the two hemispheres were not interacting at all, all reaction times should be the same. Jeff and I disagreed. We told her that the most logical explanation was that there was a cross-cueing strategy going on, one completely separate mental system cooperating with another mental system hopelessly intertwined, since both had to use the same body to express themselves. In that unique kind of situation, strategies for action would be in order. The strategy appeared to be this: Respond fast if it's a vowel. That explains why two vowels are the fastest, as each hemisphere knows this rule. It also easily explains why two consonants were the slowest. In that case, each brain was waiting to see if the other was going to opt for a fast response. When that didn't occur, each side could independently conclude that both sides must have seen a consonant. In the conflict conditions, while one side may have wanted to respond quickly, the other through various subcortical strategies was trying to go more slowly.

Jeff and I were convinced we were right, but at the time, Sergent had her own views. Within a month, she had the study published in *Nature!*[9] We sent her other data that we had collected on patient V.P. that supported our view—she saw it differently. So, we decided to disagree about it and let it go. Over the next few years, however, the view was trumpeted by Sergent and others, including several studies from the Sperry lab. While Sergent acknowledged a few years later that the study she did on J.W. was flawed, she had gone on to test some of the West Coast split-brain subjects and felt overall that her ideas were confirmed. So, seemingly out of nowhere, it was now commonplace to see the argument that higher-order information did transfer between the hemispheres but specific details of perceptual information did not. What was going on? It was time to do a thorough study, and this was a new graduate student moment for Sandra Seymour at Dartmouth. It wound up taking a few years to complete the work, but it was a beautiful and complete study.

Seymour reviewed all of the published data on all of the split-brain patients in the United States. She determined that only two patients were providing support for the so called "reunified" view of the split brain: Cases L.B. and N.G. from the California series. Case L.B. was problematic for many reasons, one of which was that it was not clear if his callosum was completely sectioned. MRI results were mixed. On cross field comparisons of perceptual information, he scored more like normal than did the split-brain patients. As a result, and before trying to understand the puzzling results from L.B. and N.G., Seymour decided to rerun the East Coast patients, J.W., V.P., and D.R., on all the tests run by Sergent. It was Sergent's work that represented some of the strongest claims that subcortical pathways were responsible for the interhemispheric integration of higher-level abstract information.[10] In fact, she argued that it is the very abstract nature of the information that made interhemispheric comparison possible. Sergent reasoned that the subcortical pathways were less efficient at, or incapable of, transfer or cross-comparison of stimulus identity. In short, she reported that performance might be compromised when simple physical identity was emphasized, whereas performance improved when the same stimuli were being compared for meaning. I still don't fully understand how Sergent came to that way of thinking.

At any rate, Seymour retested the East Coast patients using the exact tests Sergent had used on the two key Caltech patients. She also tested Sergent's proposal that abstract representations, but not sensory information, could transfer in the split brain. On this test she required the patient to compare numerical values represented by a digit in one visual field and a group of dots that added up (or not) to the digit value in the other visual field. This task could be performed only if the abstract idea—that is, the notion of, say, "seven"—were shared between the hemispheres.

In the end, we simply could not replicate the results reported by Sergent for the West Coast patients L.B., N.G., and, to some extent, A.A. So, what was going on? Indeed, several other researchers were beginning to report interhemispheric interactions testing the Caltech patients. This was even more puzzling since, as already pointed out, their surgery involved more interhemispheric disconnections than did that of the East Coast patients. Put simply, the East Coast patients appeared more disconnected, not less, even though their secondary commissure, the anterior commissure, was intact. How could this all be?

We tried everything we could think of in the perceptual realm. For example, we tried to see if V.P. could compare two simple wave patterns, one presented to each side of the visual midline. Nope, no deal. She couldn't cross-compare two nonnameable symbols, a very simple task. J.W. couldn't do this kind of task, either. And P.S. had failed to compare words or pictures many, many times.

To begin to understand the likely explanation, it should be kept in mind that the ability of split-brain patients to look as if they were integrated develops over time. Immediately after surgery and for some time, this seeming ability to transfer information across their disconnected brains is clearly not present. The effects only come from the highly practiced patients after years of testing—or living with someone for years, as I have already pointed out. When this kind of yoking happens between two people, no one is surprised or mystified.

Now, imagine conjoined twins, say, conjoined at the neck. Yes, two completely wonderful human beings coming out of the same neck, like two flowering roses coming off the same stem. Such cases exist and have been well documented.[11] The twins Abigail and Brittany Hensel, now in their twenties, who grew up on a farm in Minnesota and graduated college with teaching degrees in 2012, have two quite different personalities. They are unquestionably two separate mental entities, but also reveal the myriad of ways they cross-communicate to keep purposeful actions of their shared body, such as playing softball, integrated.[12]

Let's say the conjoining is at a higher level. Basically, the split-brain surgery has disjoined one mental system and made it two. One of the systems, the left brain, is very bright and creative; the other also has a set of skills, but they are different. Nonetheless, two distinct mental systems that were formerly joined are going to have to learn how to get along without having the direct neural communicative networks that they previously enjoyed. They are going to have to learn a lot about cueing, about nonverbal communications, about how, in fact, most humans live their days, tipping off others as to their desires, frustrations, and impending actions by subtle, very subtle, cueing. There is not too much argument about this reality.

Our argument was that this cueing skill markedly improves over time, making it appear as if the split-brain patients were reconnecting after many years of looking so different. For lack of a better phrase, we came to call these strategies part of a "readiness response."[13] Thus the positive interhemispheric results that are observed when performing value comparisons between numbers shown to one hemisphere and dot arrays of same or different amounts shown to the other can be explained by a cueing system: Each hemisphere independently, and without any knowledge of the stimulus presented to the other, has a disposition to respond in a manner that is determined by the magnitude of the digit presented to it. The hemisphere more disposed to respond then initiates the motor output. In other words, if each hemisphere decides to act if the number is high, it is possible to obtain 78 percent accuracy by simply applying a strategy based on the digit in a single visual field (that is, if the digit is <4, guess that the other side is higher; if it is >6, guess that this side is higher; and if it is 5 simply guess). We ran this test on J.W. and he hit this level of accuracy and then told us he was using this exact strategy! No communication between the hemispheres, just a cooperative strategy.

Endless variations of these experiments exist, but the point is clear: Two mental systems forced to share the same resources, work it out. It took two very bright and talented young scientists to get it all straight, Seymour and Reuter-Lorenz.

BRAIN MECHANISMS OF ATTENTION

Just as colleagues were ready, willing, and able to visit Hanover to carry out studies on our patients, we, scientists and patients, were ready to travel to far-off places for testing. This was especially true for places like Steve Hillyard's lab at the University of California, San Diego, in La Jolla, with its unique natural setting.

As I mentioned earlier, Hillyard used event-related potentials, the complex brain imaging procedure that allows one to study both the timing and, to some extent, the actual place in the brain that generates particular brain waves.[14] When he first arrived at UCSD, Steve was housed in the Scripps Institution of Oceanography, on the beach. As the years rolled by, the university built a more traditional building up on the bluffs and Steve lost his coveted space. It is there that I met his hotshot student, Steve Luck. Luck describes his first testing session with J.W., who'd accompanied us out west.

My very first experience with a split-brain patient was with J.W. in an experiment on visual search. For reasons that I still don't understand, people in the Hillyard lab thought I was fully competent to test J.W. with absolutely no prior experience. I brought him into the lab, sat him in the chamber, and explained the task.

I said something like this: "In this task, the target is a rectangle formed by a red square on top of a blue square. The distracters have a blue square on top of a red square. Your job is to press the left button if you see the target on the left side of the screen and to press the right button if you see the target on the right side of the screen. In other words, press with the left hand if you see red-on-top/blue-on-bottom on the left side and press with the right hand if you see red-on-top/blue-on-bottom on the right side."

I then asked if he understood, and he said "sure." He was obviously a pro, so I figured he understood perfectly. Well, his left hemisphere—which was the hemisphere that was talking to me—did understand the task, but this was way too syntactically complex for his right hemisphere. So when I started up the task, his right hand pressed the button every time the target appeared on the right side, but he didn't make any button presses with his left hand.

I stopped the task and went back into the chamber. I explained the task again, but all the while he protested that he understood the task. I left the chamber, started the task again, and J.W. again did perfectly with his right hand/left hemisphere but made no responses with his left hand/right hemisphere.

I stopped the task and tried explaining again. He again said he understood, and was starting to get a little peeved that this kid was trying to explain something that he, as a professional subject, clearly understood. But again his left hand did not respond.

And then I suddenly realized that I was trying to explain this complicated task verbally to a hemisphere that had limited language abilities.

I went back into the chamber, and I said "Please be patient with me while I try this one more time." I started up the task, and every time the target appeared on the left, I pointed at the target and at his left hand, saying "blue-top, left-hand, blue-top, left-hand." He kept protesting that he understood, and then suddenly a funny "aha" look came over his face. He then said, "OK. I'm sure I understand now."

I left the chamber, started up the task, and both hemispheres did perfectly from that point onward. We got some great data that led to a Nature *paper, and I learned how to explain a task to the right hemisphere of a split-brain patient—a hemisphere with poor syntactic ability.*[15]

Steve was just getting his feet wet in science, but it was already clear he would be a star. In fact, most of Hillyard's students have become scientific leaders. His standards were impeccably severe, and his insights were frequent. Before Steve Luck entered Hillyard's lab, Marta Kutas, Ron Mangun, Marty Woldorff, Bob Knight, Helen Neville, and others—all household names in neuroscience today—had come under his tutelage. Collaborating with any of them always led to solid research. Still, each was a novice when it came to studying patients. All had cut their teeth on examining normal college undergraduates, who had allowed one to talk as Luck described. It took experience to learn how to describe what you wanted to two very different disconnected hemispheres. The only way to learn was by doing it—trial by fire. As you may be coming to understand, our patients were patient, exceedingly.

Understanding human attention is one of the grand challenges of modern cognitive neuroscience. In many ways, the best and brightest researchers have been committed to various aspects of the problem. The field was beginning to understand how attention could be directed to particular points in space in order to enhance the sensory moment or how it could be directed away from one conversation to hear another conversation. Attention was thought of as a beacon of light swinging through the rich landscape of our sensory experience, focusing on the specifics of the scene we are currently engaged with. It was the great enhancer to both perception and cognition. Naturally, we began to wonder: Does each hemisphere of a split-brain patient have its own attentional system, or is it shared? Could one hemisphere attend to the left while, at the same time, the other could attend to the right side of space? If you have an intact callosum, you cannot do this at all.

Once again it was Jeff Holtzman who had laid the groundwork. In many ways, the problem of attention seemed mercurial. Either half brain of a split-brain patient was able to direct attention to places in its sensory world. What surprised us was that each half could also direct attention to specific places in the part of the sensory world it did not have direct access to, places that were in the other half brain's bailiwick. This exception to the rule, that spatial attention could flow across the disconnected brain, seemed weird.[16] So, we wondered if it were possible for each half brain to direct its attention to a different place at the same time. Would that be a nonstarter? Was it like, say, a tight end being asked to be at two different places at the same time? Apparently it was.

Patti Reuter-Lorenz nailed this important idea at Dartmouth.[17] The attentional system was unifocal. In short, the two disconnected hemispheres could not prepare for events in two spatially disparate locations. Something was still glued together in the split brain. There seemed to be some sharing going on of a common resource—for lack of a better word, let's call that resource "oomph," energy—the stuff that is drawn upon to do anything. This idea led to a further refinement of the different kinds of attention the brain calls upon to do its work.

In an earlier study that Jeff and I did at Cornell, we showed that J.W. would react faster from one hemisphere if the other hemisphere was working on an easy problem instead of a hard problem. We supposed that in order to solve a hard problem, more resources would be drawn off than would be to solve an easy problem. When the hard problem was being shown, the opposite hemisphere would, as a consequence, be slower to respond to the different task it was being asked to solve at the same time. Somehow, resources were common to both hemispheres.[18] We thought we had it confirmed.

There was this nagging feeling, however, that we hadn't fully characterized what was going on. From my Caltech days on, I had been showing that split-brain monkeys seemed able to respond accurately to more information presented in a brief flash than normal monkeys were. At one level, it seemed like the animals' resources had been expanded and improved, not lessened. Jeff and I had found a similar result with human patients. What, indeed, was going on?

This is the test we'd run: Imagine looking at a point in space or even better, a point on your laptop (Video 11). On each side of the fixated point there is a box divided into 9 cells, 3 up, 3 across, like a tic-tac-toe grid. Now imagine the experimenter is about to present to you a sequence of four X's, distributed one after the other in 4 of the 9 cells, which you are to remember. Further, this memory test will be presented in both visual fields at the same time. I am kidding you, right? No, that is what we did, and we did it in an easy way and a hard way. In the easy way, the nine-cell box in each visual field had the same sequential pattern presented, so we called that the redundant condition. In the hard condition, the box in each visual field got a different pattern of sequences. Trust me, that is a hard condition. After either the easy or the hard stimulus sequence had been presented, another pattern of four X's—a probe—came on, which either matched the pattern in the field that had just been observed or did not match the pattern in that field. All the subjects had to do was hit a button marked "yes" or "no," meaning yes the probe was identical to what it had just seen, or no it was not.

Nonsplit subjects whipped through the easy trials. They were fast and accurate. Even though there were 8 different X's coming on quickly, 4 in each visual field, they were easy to apprehend because the X's were coming on in the same sequence and in the same pattern in each visual field. They were redundant. Because of that, it is easy to do. J.W. found it easy to do, too.

The hard trials were a different matter. It stopped even smartypants undergraduates in their tracks. It was too much information presented in a brief time to grasp. Robert Bazell, the distinguished NBC science reporter, was visiting during one of our tests and exclaimed, after seeing the flurry of stimuli, "What on earth was that?" Clearly, the normal memory system couldn't handle it, and accurate responses fell to chance.

Not with J.W. When the mixed trials came along, with each hemisphere receiving 4 different X's at 4 different positions in the nine-cell box, J.W. held on to the information and kept on getting the correct answers. It was like he had two independent processors, which made for better scores when combined.[19] It looked like the common unifocal attentional system, which we thought we had captured in our previous studies, couldn't explain this remarkable increased capacity. The cool thing about science is that explanatory models, which have been proposed to explain mechanisms, can change, to the disappointment or enthusiasm of the researcher. As a scientist, you need to be flexible. If new data disprove your belief, you have to change your belief. The scientific field of attention is peppered with truly great and congenial researchers. They are happy to adjust their ideas to fit new data. Holtzman, Reuter-Lorenz, Luck, Mangun, and Kutas were about to change split-brain research.

Having these young experts and, of course, one of the doyens of attention research, Hillyard, on the case was, as they say in Texas, high cotton. From another angle, it was plenty as well. When running a broad-gauged lab, with only so many slots, only so much money, and with so many issues to be studied, it becomes necessary to limit who else joins in the effort. Those are the plans and that is the management theory. Then Alan Kingstone walked into my life, a Canadian student of another famous attention researcher, Ray Klein at Dalhousie University in Nova Scotia. Great, I mumbled to myself. I need this like the proverbial hole in the head. Kingstone warmly tells the story like this:

> . . . *Michael Posner pointed me towards Michael Gazzaniga.* . . .
>
> *So in the blissful ignorance bestowed upon the very young, and those that hold Ph.D.s, I picked up the telephone, used my dime, and called Michael Gazzaniga. He answered, and deduced in about one millisecond that I didn't know anything about the brain and its relation to human cognition. The conversation went something like this:*
>
>> *Mike: Do you know anything about the brain? [This, I was to learn, is classic Michael; he cuts straight to the heart of an issue, or as was often the case where I was concerned, the weakness of an issue.]*
>> *Alan: No. [This was not going the way I had hoped!]*
>> *Mike: Don't you think that's a problem?*
>> *Alan: No. I'll learn.*
>> *Mike: Come on down. Let's see what you've got to offer.*
>
> *Truly. That was it. A short time later I found myself flying to Montreal, and then catching a train on a beautiful spring morn-*

ing through the magical countryside of Vermont down into White River Junction. From there it's just a ten-minute cab ride to Dartmouth College, in Hanover, New Hampshire. And by the time I stepped out of the cab and onto that remarkable Ivy League Dartmouth Campus—a campus that manages to blend the old and new together in such a seamless manner—I was completely and absolutely sold. And I began to suspect the truth—that my life had already begun to change forever, and for the better.

The next day I set off for Mike's lab. At that time, in the early 1990s, Mike and his team were conducting their research in a white clapboard, side-gabled house that had been built by Mrs. A. Pike in 1874. There at Pike House I met several of the future stars in cognitive neuroscience: people like Patti Reuter-Lorenz and Ron Mangun, and of course, Michael Gazzaniga himself. He had me give a little talk to his group, and then whisked me off to an elegant French restaurant, where he offered me a spot in his lab. "Say yes, and come do your thing" was his offer. I accepted of course. We shook hands, and that was that.

It really only took about three minutes to decide to hire him and it only took him two seconds to accept the offer. He had that glint in his eyes, the energy level of a buzz saw, and the smarts of the rest of the group. He also had a thirst for new problems, new angles, new adventures. So I changed my theory on the spot and decided we needed to dig into attention issues more deeply.

ATTENTION REDUX

Steve Luck was busy in San Diego puzzling about what split-brain patients were capable of doing and how they did it. He carried out an amazingly clever experiment that, at one level, confirmed the idea that they were capable of enhanced information-processing capacity. How could that be? Luck went after the problem by applying a well-established test from the experimental attention literature. He took an array of blue and red squares that were stuck together with the blue square on top and spread a bunch of them out on a computer display screen. Each time he presented the array of squares, he snuck in one square pair that was different. It had the red square on top while the yoked blue square was on the bottom. The squares in the array were called the distracters, and the single red/blue square was called the target. The task was simple: Find the target.

When neurologically intact subjects do this task, an interesting and consistent behavior occurs. As more distracters are added, it takes longer to find the target. In fact, our response time goes up in a reliable way. Every time two more distracter squares are added to the distracter array, it takes another 70 milliseconds to respond. The distracters slow down our search to find the one target. This happens like clockwork. It also doesn't matter where in the left or right visual fields the added distracters appear.

Split-brain patients respond in a dramatically different manner. When the extra distracters are added to a single visual field, the patients, not surprisingly, take longer to find the target, like everyone else. However, when that same number of added distracters is spread out such that each field gets half of them, the overall reaction time is much faster when compared to everyone else. In other words, each disconnected hemisphere seems to have its own attentional scanning machinery, and each can go to work simultaneously and independently of the other half brain. Luck did these studies on J.W. and also on the Caltech patient L.B.

This was an exciting finding, which was well documented and robust. It was beginning to look like there were many components to the attention system. It looked like some aspects of attention were involved with scanning a visual scene for particular information. Other parts, which were associated with cognitive work, were still connected, presumably through lower brain systems.

Kingstone pushed on these ideas and made them even more interesting. He wondered if each hemisphere was doing its scanning of the arrays using the same kind of strategy. After all, the left hemisphere was the smart, verbal hemisphere, while the right was specialized for grouping visual parts into sensible wholes. Maybe their underlying attentional mechanisms served up their discoveries of the visual world by different means. Alan made the target selection process more difficult. He added even more distracters, such that when using the low-level automatic systems described above, we humans begin to crack. We are smart animals, so we guide our attention through cognitive strategies. In a word, we start using "top-down," that is, goal-directed and "guided" ways to sort through all the information. Let's say the task is, Find Louise. One strategy might be, "Look for big hair!" Alan discovered that we can only do that in the left hemisphere.[20] The right hemisphere is stuck doing the searches in the standard automatic way. "Look at everyone until you find Louise." All of this work led us back to an even stronger view about split-brain patients: Not only were the half brains separate, those puppies were also different!

BRAINPRINTS AND THE SWISS CONNECTION

The lab was bubbling with activity. In part this was because it is not in my nature to focus on one topic. When I was growing up seventy years ago, there was no such disease as attention deficit/hyperactivity disorder, so I couldn't have had it! Now, as I look back, I wonder. My mother always said that I had ants in my pants. Drilling deep on only one topic is a popular way to spend one's life. But it's not for me. When working with patients and the ephemeral topics of attention, which requires from hundreds to thousands of boring trials of having patients react to simple lights, my mind became inattentive. I wanted to do different things, more closely connected to basic neuroscience.

We had settled into Pike House, but we were bursting at the seams. If we were going to expand, how? Pike House had an outdoor porch. I asked if the medical school would enclose it for additional office space. I called the contractor who had built my new house, had him give a bid, and the provost quickly agreed to it all. Over the years I had learned that if you present a problem to an administrator, also present the solution. Then it's only money, and usually administrators can handle small budget items. The job was finished so quickly (that's how you actually make money in construction) that the college took a shine to my builder, Rusty Estes, and used him frequently thereafter.

Soon, however, we were once again overcrowded. More grants were rolling in, postdocs were flocking, and our new graduate program was filling up with students. The medical school started to get more interested in our enterprise and moved us over to some space in the actual medical school building. The walls were god-awful yellow tiles only matched in ugliness by the well-used linoleum floors. That actually was the first offer. I said I wasn't leaving our beloved Pike House for that. Okay, said the dean and, after a new paint job and new carpet, over we went. It turned out to be a delightful space and further energized the lab. We could now expand the lab's pet project: mapping the human brain.

As I already mentioned, my appointment was in psychiatry. For bureaucratic reasons, a Ph.D. could have tenure in a medical school department, and psychiatry filled the bill. My professional associations were with the neurologists and, in particular, the neurosurgeons. When I met David Roberts, who is now the chief of neurosurgery at Dartmouth, he was a resident under the neurosurgeon Donald Wilson, who launched the Dartmouth split-brain series. When Wilson tragically died from throat cancer, Dave took over his lead and is now the world's authority on split-brain surgery, even though it is rarely done these days.

Dave was also a Princeton man. A few years later, I was visiting Princeton on a short sabbatical. My host, George Miller, who had moved from Rockefeller University, suggested we have Dave down to talk about MRI-guided microscopes for neurosurgery operations. It was the dead of winter, but Princeton had called, so Dave answered. He got on the tiny airplane that flies out of Lebanon, New Hampshire, and showed up to talk to the Psychology Department about his work on the microscope. All I can tell you is that it was one of the best talks I have ever heard (and I have heard *a lot* of talks). The audience of nonsurgeons was mesmerized. For a neurosurgeon, one of the challenges is that although a brain tumor may be visible on an MRI, there is still the problem of finding it in the 3-D brain during surgery. Dave's form of brain mapping, using his MRI-guided microscopes to solve this problem, was riveting.

Back at the lab, we had been working on our own brain mapping project. We had started it back at Cornell as a passion of Marc Jouandet's, a graduate student of mine who joined up with the lab during our Stony Brook days. Marc was an extraordinary talent, driven and smart. He started out building the lab a computer (in a suitcase, which, in those days, deemed it portable) that could help with our studies. He picked up the parts at Tandy Corporation, aka Radio Shack, figured it all out, and bingo, we were into a form, a very early form, of data processing. As it turned out, though, Marc was an anatomist at heart. He came up with the idea of brain mapping, which we enthusiastically called "brainprints," sort of like fingerprints. We all were going to have our own unique brainprint: another easy-to-say, hard-to-do idea.

Marc always saw the possibilities in both science and life more generally. At one point during my time at Cornell, Charlotte and I planned to take a sabbatical leave so that I could write my first book for general readers, *The Social Brain*. Most sabbaticals last for a year, a time allotment that did not work for me, due to the needs of my family and running a complex lab. I just couldn't be away for such a long stretch of time. I hit upon the idea of splitting it up. Travel to a place for a month, then return home, then travel again. I mentioned this in a letter to Marc. Before I knew it, Marc, who had carried out a postdoctoral fellowship at the University of Lausanne, had found the perfect Swiss mountain cabin for us in the village of Caux, just a cog-train ride up the mountain from Montreux—a three-bedroom chalet for $150 a month. We not only booked it; we found ourselves returning to Caux for a month each winter for years.

At Caux, we skied, we worked, we hosted visitors, and we also had the yearly treat of visiting Bill Buckley around the bend in the mountain at Rougemont, where he and his wife wintered each year. During the ten weeks that Bill spent in Rougemont, he managed to write an entire book, maintain his three columns a week, continue to edit *National Review,* and ski every afternoon after lunch at the Eagle's Nest.

Bill once complained that Swissair had a rule about flying with dogs, which he found frustrating. There could only be one dog per cabin class and there were only three classes. The Buckleys had three dogs. This meant that they could not sit together for the flight. Pat, his wife, sat in first class with one dog, Bill in business class with the second dog, and the housekeeper in coach with the third dog. For years Bill tried everything he could think of to get this federal rule waived. Nothing worked. Ah, I thought to myself—maybe my brain science connections could come to the rescue of this social inconvenience. I mentioned something to Bill, and he looked at me with that "yeah, right" face. The matter was dropped.

As the next season approached, I remembered this story and decided to call up Liana Bolis, a neuroscientist and benefactor to the field in dozens of ways. I had written a monograph for her foundation and had attended several superior workshops she had organized, including a trip to China as part of a World Health Organization team to examine neuroscience in China. She was a significant contributor to the Catholic Church in Beijing, which once bound us to listen to a Chinese opera for five hours. More pointedly, she had a significant stake in Swissair. I called her to say I had this special American friend who traveled to Zurich all the time from New York and . . ." She said that she had little to do with the operations of the company but that she would get a manager to call me.

In Swiss fashion, the call came through rather quickly. The manager was very polite and amused by my request. I mean couldn't they, just once, let the Buckleys both sit in either first or business class? He thanked me for my concern for my friend and after some vague remark we both let the conversation end.

The Buckleys' flight took off two weeks later. The following night I got a call from Bill. "Mike, last night at JFK, just before the plane took off, the Swissair steward came over to me and asked me to leave my dog in my business class seat, said that he would be fine there and escorted me to an empty first class seat next to Pat! You accomplished what the entire social structure of Gstaad could not and that includes Roger Moore [of James Bond fame]." We all had a good chuckle, but I knew brain scientists had just gone up a notch in his estimation.

At any rate, by the time we arrived at Dartmouth, the brain mapping project had attracted lots of people in the lab, such as our very own neurologist, Mark Tramo. The idea was to take a patient's MRI scans in cross section and to ultimately have a computer program automatically read the hundreds of scans and generate, from that data, a flat map of the brain, which could be comprehended, visualized, and measured more easily than a real three-dimensional brain. What surprised us was that one of the psychiatrists, Ron Green, skilled in treating serious mental disorders, also took a shine to the project. He started committing hours to the tedious task of tracing the cross sections. At that time, automatic tracing was not available; hand tracing on tissue paper laid over the brain scans by skilled neuroanatomists had to do. It is a beautiful thing to see people who, although they have day jobs, are charged up by an idea and will work endless additional hours on it. This method went on for years until a bright undergraduate from Cornell, William Loftus, came along. He began to figure out how to do it all on a computer. The U.S. Navy's Office of Naval Research bought us a fancy computer, and Loftus harnessed it for the work.

Techniques are important in science. What's more important, however, is using them on important questions. Before developing the brainprint, we had determined from MRI scans that the corpus callosums of identical twins were more similar to each other than to unrelated controls. This was one of the first demonstrations that actual brain structures in identical twins were more similar than not.[21] With brainprints, we wanted to extend that idea by carefully mapping the surface of the cortex to see if other specific regions of the brain were more similar in twins. We tried, but we were not able to capture it. In the end, our brainprinting process was too laborious, and our subject pool was too thin. That, however, doesn't mean the question wasn't addressed by others. Discovering the similarities and differences of twin brains was taken up by the UCLA brain imaging group, and they firmly established how similar twin brains are, both structurally and functionally, using sophisticated and advanced brain imaging techniques that far surpassed ours.[22] Still, this experience prepared us for a different "big science" project a few years later. The seeds were laid, but it took about eight years for them to bloom.

ONLY PARTIAL DISCONNECTIONS: THE SEMI-SPLIT MIND

In some sense the overall goal of biological research is to strive to make observations more and more specific. The first success of showing the dramatic effect of a full callosal-splitting surgery, where basically nothing seemed to cross over between the two half brains, soon gave way to the question: What if only specific parts of the callosum are sectioned? Or, what if specific regions remain after surgery? Both of these issues were always on our minds, and opportunities to study such problems popped up unexpectedly.

Classical anatomy of the callosum indicates that the posterior regions of the structure interconnect the visual areas at the back of the brain. As one moves forward, the fibers connecting the parts of the cortex responsible for hearing, touch, and other body sensations and movement become evident. Armed with this knowledge, one would predict that a lesion to the posterior regions of the callosum might cause a problem with the transferring of visual information between the hemispheres. In a way, the idea was that one might see a "modality-specific" split. That is, such a patient might be visually split but not split when tested in other modalities.

Years earlier, I was sitting in my office at NYU when a Brooklyn neurologist called me about a couple of cases. He was following two patients who had had their posterior callosum sectioned as a consequence of a neurosurgical procedure to get at a tumor of the third ventricle, a place in the brain located just below the posterior callosum. He asked if I would like to study the patients. After leaping out of my chair with excitement, all was arranged and led ultimately to a paper that we jointly authored. I love this part of a life in science. A practicing neurologist, though a stranger to me, keeps up with the literature, sees some patients in his office who might be of interest to a basic researcher, discusses it with his patient, who agrees, takes the time to find the researcher (in the days before the Internet), and then, importantly, participates in a research effort. Who says we are not an altruistic species?

These two patients taught us many things. The first case was visually split, just as predicted. His other modalities had been left intact. (He also turned out to be one of the minority of people in whom hemisphere dominance is reversed.[23] It was clear from the pattern of his responses that his right hemisphere was dominant for language and speech, while his left was dominant for the usual right hemisphere specialization, such as drawing in three dimensions.) One case, one random call, and we were closer to understanding how the parts of the callosum were organized.

Over the years, other cases were brought to our attention by clinicians, and they too provided more insight into the organization of the corpus callosum. For example, another patient, E.B., had a slightly more extensive posterior split. As expected from the known anatomy, it seemed to prevent tactile and auditory integration. She also had a remarkable ability to integrate motor information in one direction, from the left to the right brain but not from the right to the left brain, again suggesting great specificity of connections.[24] After all, where the surgeon actually stops sectioning the callosum is somewhat arbitrary. It makes sense that what information systems get disconnected should vary. These clinical cases, which were presented to us through indirect routes, were enormously interesting. They were made more so because, as a result of the main research program on split-brain patients, we knew what questions to ask.

Still, it was two of our star patients who truly illuminated a few secrets of the callosum. J.W.'s callosum had been surgically split in two stages while we were still at Cornell. His posterior callosum was sectioned first. Ten weeks went by before he underwent his second anterior section. This gave us the unique opportunity to examine him before his surgery, and then after each successive surgery. Preoperatively he was completely normal with respect to our tests. The two half brains were in total communication. John Sidtis, who was another one of my ace postdoctoral fellows, Jeff, and I examined him again after his posterior callosum had been sectioned.

Wilson stopped the surgical section approximately halfway along the callosum, a little farther toward the anterior regions of the callosum than the sections of either of the two clinical patients I just described. According to our standard tests, which carefully examined each modality, J.W. seemed completely disconnected. While that was exciting, we knew that the entire anterior half of his callosum was still intact. Since the posterior half section seemed to result in the full split-brain syndrome (as we then understood it), we wondered: What on earth is the anterior portion transferring? What were those 100 million or so neurons in the front of the callosum doing? Sidtis and Holtzman kept pushing.

After we did the routine test of flashing simple pictures to each visual field and determined that J.W. could easily name pictures flashed to the left brain but not pictures flashed to the right brain, we

242 of 452 | TALES FROM BOTH SIDES OF THE BRAIN

wondered if he could carry out some other kind of cross-integration of information. We set it up by flashing a stimulus to each field. The left hemisphere saw the word *sun* and the right hemisphere saw a picture of a simple black-and-white line drawing of a traffic light. Our simple question to J.W. was "What did you see?" The conversation went like this (Video 12):

> MSG: What did you see?
>
> J.W.: The word *sun* on the right and a picture of something on the left. I don't know what it is but I can't say it. I wanna but I can't. I don't know what it is.
>
> MSG: What does it have to do with?
>
> J.W.: I can't tell you that, either. It was the word *sun* on the right and a picture of something on the left. . . . I can't think of what it is. I can see it right in my eyes and I can't say it.
>
> MSG: Does it have to do with airplanes?
>
> J.W.: No.
>
> MSG: Does it have to do with cars?
>
> J.W.: Yeah (nodding his head). I think so . . . it's a tool or something . . . I dunno what it is and I can't say it. It's terrible.
>
> MSG: . . . are colors involved in it?
>
> J.W.: Yeah, red, yellow . . . traffic light?
>
> MSG: You got it.[25]

The anterior callosum was beginning to yield its secrets. Somehow the right hemisphere was passing forward to its more cognitive parts what we thought were the more abstract aspects of the line drawing. Somehow the various associations yoked to a picture of a black-and-white drawing of a traffic light were activated, and the parts of the brain supporting those more forward-based functions still had their callosal connections. These associations were prompted by the game of twenty questions I was playing with J.W., the game that was being managed by the left hemisphere. The anterior part of the callosum dealt with higher-order information, not the primitives of the actual stimulus. Again, it wasn't a representation of the actual image: It was other gnostic associations that the left hemisphere was receiving from the right and trying to find words for.

Still, one could argue that nothing was actually being transmitted across the callosum. Instead, maybe the left hemisphere was simply being cued by the isolated right hemisphere in this fashion: When I ask, "Does it have to do with cars?," the right hemisphere hears "cars," which are associated with a traffic light, and so nods *yes*. The left hemisphere notes that the head nods *yes* and goes along with the cue and says "yes." Then I ask, "Are colors involved?" and again the right hemisphere associates colors with the traffic light and once again nods *yes*. Now the left hemisphere knows: "cars" and "colors." Quickly, all on its own, it figures out what the picture to the opposite hemisphere must have been, just as anyone might have done by listening to the exchange. This argument proposes then that nothing actually transferred through the anterior callosum. Instead, perhaps, utterly separate and independent modules are cueing each other, just like two people cue each other with the wink of an eye.

But things kept changing with J.W. The traffic light account came early on after his first surgery. As time elapsed, we saw that J.W. could play the "twenty questions" test with himself without outside direction. Another change came about eight weeks into the period between the first and second surgeries. We flashed the word *Knight* to the right hemisphere. Here is the dialogue he had with himself: "I have a picture in my mind but I can't say it. Two fighters in a ring . . . ancient, wearing uniforms and helmets . . . on horses trying to knock each other off . . . Knights?" The word *Knight* elicited all of these higher-order associations in the right hemisphere.

These were being communicated externally through speech and hearing over to the left, which picked up the parts and then solved the problem.[26] This was extraordinary and made even more so by the fact that following his second surgery, which completed the full section of the callosum, he could never again succeed in naming words and pictures presented to the right hemisphere. At least he couldn't until something else changed, but that comes later (chapter 7).

ONLY THE LEFT BRAIN SMILES TO COMMAND

The patients were and are endlessly fascinating, and our testing program was incessant. Everyone knew each other, and the patients were as supportive of us as we were of them. My wife, Charlotte, was a big part of the glue that kept it all working. Patients would ask for her even when it wasn't one of her days to test. On those days, Charlotte would still be part of the group that took a patient out to dinner. Or, if there was a birthday for a patient's child, Charlotte would remember, and a present would arrive at the lab for the patient to take home. Charlotte's naturally hospitable Texas ways were always there. And just as the patients needed to feel at home and comfortable, so did the visiting scientists. I can't count the number of dinners Charlotte cooked, and all of this while making use of her own training in neuropsychology and doing her own experiments. The social aspects of a life in science are considerable and extremely important if the science deals with people. In the days of the traveling van, she somehow would transform the testing area into a dining room and serve a four-course dinner that miraculously emerged from the small oven and stovetop! It was—like it sounds—magical, as is she.

Charlotte had been reading up on a curious fact about the anatomy associated with voluntary versus involuntary smiling. The brain has allocated these two very different skills to different brain systems. When you smile voluntarily, when you are asked to smile, the act is controlled from the left hemisphere. It involves the cortical neurons that cross over to the right half of the face and also the cortical neurons that cross through the callosum. There they activate other cortical neurons that ultimately activate the left half of the face. This all happens very quickly, such that when that smile comes out, it looks perfectly symmetrical. If a stroke has damaged any part of this cortical pathway's network, however, there can be a corresponding droop in the smile, depending upon where the lesion occurs.

Spontaneous smiles are different. They utilize a totally different neurologic hardware that is diffuse and arises mostly out of the subcortex and something called the extra-pyramidal system. When you hear a good joke, this is the system that kicks in and produces the giggly face. Why is it that gramps, who suffers from Parkinson's disease, looks so deadpan? Because his disease damages that extra-pyramidal system, with the unfortunate result that he can no longer smile spontaneously.

Charlotte reasoned that our patients should reveal this, if we tested the idea properly. We knew how to ask the question: simply flash a command to either the left or right hemisphere and record on video the response. By pointing a video camera right at the face, we thought we should be able to pick up a possible difference in which part of the face first responded. A flash of the command to smile to the left hemisphere ought to find the right side of the face commencing the smile, followed by the left side. Or a flash of the command to the right hemisphere should find the opposite, the left half of the face moving first, if the right hemisphere could carry out the command. Sounds easy but, of course, there was a hitch. The video camera we had at the time couldn't capture the frames fast enough to see these split-second differences.

I had been toying with the idea of buying a Panasonic digital videodisc recorder. Several other projects in the lab, including our brainprinting project, needed a way to store large amounts of data.

This video camera could not only do that, it could also capture information at a much faster frame speed and could play it back frame by frame. Would this do the trick?

Charlotte and I hooked it all up and started testing J.W., V.P., and D.R. It worked perfectly. Take J.W. as an example. Sure enough, when the command to smile was projected to the left hemisphere, his right half face led the smile, followed quickly by the left (Video 13). It was amazing to see, and we were anxious to see what the right hemisphere could do. To our enormous surprise, the right hemisphere couldn't carry out the command, period. Initiating a voluntary smile was not an option for the right hemisphere.[27] Yet it had no problem following the commands "wink" and "blow." At the same time, the patients had no problem spontaneously smiling to a joke or other natural situation. The subcortical control system had not been affected by the split-brain surgery.

THE ALLURE OF A RESEARCH UNIVERSITY

It was a time in my career when I wondered about a role in leading a larger effort, something much bigger than my own lab. Johns Hopkins had been on the hunt for someone to lead a new mind/brain program they were initiating. After a few visits and late-night phone calls, it didn't work out. In the end, we had different ideas about who should be hired as part of the new effort. I knew at the time that if I started to propose specific names, there would be a reaction. There always is. I had held off revealing my list of candidates until one night the chair of the committee found me by phone in a hotel in Los Alamos, New Mexico. We were down to the wire, but I reiterated my reluctance to name specific people. While I do act independently, I always consult with my colleagues beforehand. I told the chair that this was what I would do. He persisted, and, finally, I mentioned a few names. He thanked me and that was that. I never heard from them again.

Still, this had triggered in me an arousal for bigger things. When

I finally decided to take an offer from the University of California, Davis, I was ambivalent, not surprising given our life in a beautiful place with the rich and vibrant group of colleagues we enjoyed. When the decision was coming to a head, I remember standing at a phone booth at the Society of Neuroscience meeting in New Orleans and calling to check one last time with the provost of Dartmouth, John Strohbehn. There are always counteroffers in the academic world. My request was that the Dartmouth Medical School guarantee all of my salary, not just 50 percent of it. Rough waters were ahead and that seemed prudent as UC Davis was offering a full salary and more of it.

It came down to a $25,000 difference between the two institutions. That's it. If the provost had thrown in another 25K toward my salary support, I would have stayed at Dartmouth. For administrative reasons, too difficult and too unimportant to review, he couldn't do it. Strohbehn was, at heart, a terrific provost, a bioengineer who worked with Dave Roberts on the MRI-driven microscope. He wanted to make it work. I thanked him for the call and his efforts and hung up the phone and stared at the ground for a good five minutes. Okay, I thought, that's it. We are moving to Davis.

I called Charlotte and let her know. She was supportive as always, yet I could detect the strain. We had lived in our personally designed and built home for only two years. The pinewood-framed windows with hardwood floors and brick fireplaces, along with the ten acres of Vermont woods, were all going to be history. We had started many traditions in our home, the most notable of which was our dinner parties for both family and visiting scientists. Our dining room was a place of joy and intellect. Were we really leaving all of this for California's Central Valley? In an effort to soften the blow, I booked a suite at the Auberge du Soleil in Napa, packed up the family, and took off for a quick visit to show how sweet life in California would be.

The Auberge was everything it was cracked up to be. The Mexican-tiled suites had built-in sofas populated with large pink pillows that were not actually pink but some new designer color that made us all feel hip. In the dead of winter we had drinks by the pool, even though it was a bit cool in the January week we visited. Dinners over at Tre Vigna in St. Helena were sublime, as were visits to various vineyards. There was only one problem. Napa is not Davis! I should not have shown the family Olympus first!

It all worked out gloriously in the end, even with the difficulties of buying a new home when the old home hadn't sold. Throughout the process, and I do mean process, Dartmouth kept smiling benevolently upon us. It is disruptive to move, not only for a family, but also for the institutions involved. The one you are leaving is hopefully not happy about it. But when they are not happy, what do they do? In Dartmouth's case, they threw us a big party and everybody showed up to wish us well, including President James O. Freedman, the provost, and the dean. We were flabbergasted and touched, and, though we were leaving, our attachment to the institution was reinforced.

PART 3

EVOLUTION
AND
INTEGRATION

CHAPTER 7

THE RIGHT BRAIN
HAS SOMETHING TO SAY

We are at the very beginning of time for the human race. It is not unreasonable that we grapple with problems. But there are tens of thousands of years in the future. Our responsibility is to do what we can, learn what we can, improve the solutions, and pass them on.

—RICHARD P. FEYNMAN

MY MOVE TO THE UNIVERSITY OF CALIFORNIA, DAVIS, IN 1992 began when I met the neuroscientist Leo Chalupa in 1988 at one of my small Moorea-type meetings, which by now I was organizing more frequently. The topic was human brain evolution, and several of the world's experts were there. Since I only offered a thousand dollars for expenses, the place always had to be the draw. The beckoning place that year was the dreamiest city on earth: Venice. Leo was in full form.

Arrangements had been made to stay at La Fenice et Des Artistes, a wonderful small and informal hotel close to the famous Teatro. As a favor, my close and longtime friend Emilio Bizzi, a professor at MIT and a foundational scientist for understanding how the brain carries out action, had arranged our meeting place. It was across the street from the hotel at the fabled Ateneo Veneto, which had been inaugurated by Napoleon in 1810 to promote science, arts, and literature. We took over the third-floor library and made arrangements with a local bar/tabacchi to bring us espressos at regular intervals. The meeting sparkled with the likes of Stephen Jay Gould, Terry Sejnowski, a wizard in the new field of neural networks, and regulars such as Jon Kaas and Gary Lynch. The new entrant to the group, Leo Chalupa, is a man who could slide in and have tea with the queen of England at four and leave for a martini with his pals at six. Leo, with his engrained and endearing New York manner and prosody, knows who he is and also, in a profound way, knows who others are. As a mutual friend once said to me, "Leo is the guy who is there for you when others have dumped you on your head." In addition to his capacity for deep relationships, he is also very funny.

The meeting had been rolling along for a few days when the time came for Leo to speak. Several of the speakers had mentioned Francis Crick, as in "my buddy Francis Crick." This is sort of the equivalent of saying, "I am working on this idea, and I have tried it out on Crick. He likes it, so don't give me too much grief about it." This sort of stuff doesn't fly with Leo. Wanting to get his disapproval across without offending anyone, he started off with an old Yiddish joke his father told him when he was six.

> *Two guys are walking down the street—neither one of whom is Francis Crick—and one guy says to the other: "I have a riddle for you to test how smart you are." The other guy says: "Okay, what's the riddle?"*
>
> *"What's green, is up in the tree, and it sings."*
> *"That's easy, it's a bird of some type."*
> *"Nope, it's definitely not a bird."*

After many futile guesses, the second guy says: "Okay. I give up. What's green, in a tree, and sings?"

The first guy says: "Simple. It's a herring."

"A herring! How could it be a herring. Herrings aren't green."

"Well, someone painted it green."

"But herrings aren't found in trees."

"Well, someone threw it up there."

"Okay. But herrings don't sing!"

"You're right. I added that just to make it harder for a smart guy like you to figure out the answer."

After the Crick line, I am not sure we heard the rest of the joke. Leo successfully broke the heaviness of what I call "endless consideration," and the meeting sprang open and became both animated and more interesting. I wanted to know more about Leo.

As it turned out, a new Center for Neuroscience at UC Davis was being planned, and Leo was chair of the project and part of the search committee for a new director. It was all on paper when we first started talking about it. I was happily settled at Dartmouth and life was good. As time rolls by, however, stuff begins to happen. Leo and I began to run into each other at more scientific meetings and to have more dinners and lunches, "off label," so to speak. At the huge meetings, such as the annual meeting for the Society for Neuroscience, 20,000–35,000 scientists show up from all around the world. People make a huge effort to meet friends and acquaintances there, but, for the most part, not at the hundreds of individual lectures. Lunches and dinners become heavily booked, during which most of the business gets done.

The idea behind the new center was to build a point of contact and collaboration for the bustling young field of neuroscience. Leo had asked for ten positions to get it going, and he was seeking a new director, who would also oversee the hiring. There would be new space, new positions, and new funds to get the new hires going. If all the commitments were added up the way private institutions calculate such commitments, it was around a $25 million package. It was indeed a moment to try "big science," at least in 1992. Leo offered that moment to me.

During the visits to Davis to check out the town, the offer, the very idea of the job itself, I met all kinds of new people: the other faculty, the students, and most important for a task like this, the administrators. "Who would be my boss?" I would ask. The dean of biology. "Who would be his boss?" The vice chancellor. "And his boss?" That would be the chancellor. I told Leo, "Better meet them all." Leo arranged the meetings.

Davis had some of the best administrators I have ever dealt with in university life: Their word was as good as gold. That was important because in a big organization like the University of California, income streams and assets are totally hidden from simple view. There is no account number you look up to see the actual funds being allocated to your project. Funds for new positions come from one kind of source, funds for start-ups come from another source, funds for new space yet another source, and so on. In a state school, the numbers for each of those categories are constantly moving, ebbing and flowing with the needs of a huge organization. This is why trust is so important, and no one was more trustworthy than Robert D. Grey, the dean of biology at the time of my negotiations. He later became the vice chancellor. A deep believer in land grant universities, he was Kansas born, Scotch sipping, and steady as a rock. He believed in the project and adopted me. His greatest gift was that he took risks. He was not terrified of his faculty. And he went the extra mile on the little stuff, such as convincing his boss, Chancellor Theodore Hullar, to interrupt his Vermont vacation and drop by our home in Norwich. Not surprisingly, I took the job.

Of course, any suggestion that decisions about moving are wonderfully and completely rational, singular, and decisive in nature is just plain wrong. The emotions are churning while the mind is trying to figure out a logical explanation for uprooting the family from friends and a beautiful home. The friends where you currently work are also pitching for you to stay. I can remember a trip my wife and I made to New York City to visit Stan Schachter, the famous social psychologist. He, among other things, had once insisted on being placed in a four-bed hospital room instead of a single room so that he could watch social interactions when people were ill. Stan poured the drinks and sat us down in his prewar Upper West Side apartment on Ninety-Fourth Street. When in the throes of decisions like this, you are like a cocked gun, ready to go off at any moment about the dilemma to move or not. By the time we had gotten to Stan, the universities had pretty much evened up the competing offers. We were now dealing with the intangibles.

My wife was telling Stan how she had carefully explored Davis. She had gone into the parks and talked to the mothers strolling with their kids. She had talked about the schools, about the weather, about everything mothers want to know about a place. She reported all loved living in Davis. Stan looked at her and said, "Charlotte, you need to interview the people who left Davis." It didn't help matters that Stan never understood why anybody would live west of the Hudson River. After another drink or so, we reminded Stan what our mutual friend Leon use to say. "Well, you can't control for everything." Despite Stan's skepticism, we decided to take the job.

BUILDING FROM SCRATCH: JUST DO IT

Having already established trust in those in charge at the university turned out to be a good thing. Upon hitting the deck at Davis, I discovered that no real permanent space would be available for ten years, at least. The animal researchers who were to be a big part of the new program would do their work at the distant Primate Center. Others would be scattered around campus in various office/labs. This was hardly ideal.

I took to driving around Davis to learn the town, find a house, and look for space for the center. One day I spotted a building shell for sale in a research park (Figure 35). It was a handsome structure with a price tag of around $3 million. After peeking in all the windows, I found a phone and called Bob Grey. By this time he was provost. I said, "Bob, I found a building in Research Park for sale for three million. Can you buy it?" He put me on hold and then came back on the line, "Yes, we can; send the information to me." And that was that. The center now had a home, but it would take time to design the inside space and have it built, lots of time, since the city inspectors and the UC inspectors were both involved. Where would we work while all of that was going on?

I drove around some more and found another building, almost across the street from the new center. It could be rented and was ready to go with plenty of interim space. Another call to Grey and another yes. The new group would be placed in an empty building in Research Park only loosely associated with the university. Not a

FIGURE 35. The empty and unfinished building I spotted in a research park outside the University of California, Davis. I called the provost and after a minute or two, he said the university would buy it.

warm and fuzzy academic setting. It was a concrete tilt-up building with a surrounding parking lot. But we would be able to work closely together.

There was also the private move. How would we be able to buy a new house, when our old house still had not sold? It was 1992, house prices were dropping, and when they drop, they drop everywhere. A good deal in Davis meant a bad deal in Vermont! Because our beautiful new house had not sold, we were left with no option but to rent it out. What a mess.

I was sitting in some kind of meeting at the NIH, reading the *New York Times* at the back of the room. Years earlier, my friend Gary Lynch had started a company called Cortex. During one of our small meetings in Venice, we had been enjoying a late evening in the Piazza San Marco when Gary told me about his new company. Casually he said I should be on his board. In a gesture to humor him, I said sure and bought 25,000 shares for $25 total and promptly forgot all about it. As I was cruising through the business page of the *Times* that morning, I noticed Cortex had just received a large contract for $14 million. I looked up the stock and saw it was selling for about five bucks a share! I bolted out of my seat and went into the hallway and called the company. I asked their finance officer if my shares could be sold. I assumed there would be some kind of restriction on this, since I had been a board member. He told me that since I was no longer on the board and three years or so had passed, I could sell them. I hung up the phone, called my broker, and told him to sell them all. He did. A few days later a check for $100,000 arrived. We were now ready to buy a house in Davis. The joys of sheer luck are a big part of life, and to this day I buy Gary dinner whenever we manage to see each other.

Davis is really sort of a tomato patch, a flat piece of farmland accompanied by a tomato sauce factory. It is extremely hot in the summer but wonderfully balmy at night as the cool air moves in up the Sacramento Delta. The town had its origins in agriculture, and UC Davis was the research engine of the vast California agriculture industry. Agriculture needs fertilizer, and fertilizer owners build wonderful homes. With Gary's money, we bought a dazzling one designed in southwestern style with floor-to-ceiling windows looking out on a huge swimming pool, which in turn was lined with palm trees and pink oleander hedges. The front yard was a cactus garden. It looked positively Mediterranean. The whole house could not have been more different than what we'd had in Vermont, and it had one very powerful dimension: the pool. Almost instantly upon arriving, our kids, after a rather sentimental departure, seemed to forget about Vermont.

Thirty-three million people can't be wrong. California is spectacularly different at every turn. Forty minutes west of Davis are the vineyards and restaurants of Napa. Sixty minutes south is San Francisco and one hundred minutes northeast is Lake Tahoe. It wasn't long before we were totally infected by Tahoe, which led us to buy a cabin. The skiing, of course, was fabulous, but the summers in Tahoe rang sublime. And Davis had something that the Texan in Charlotte missed: a big sky. All in all, Davis proved to be a magical place to live.

The serious part of any project centers on the people who are hired. The University of California is perhaps the top-ranked research university in the world. It demands that all hires meet the highest standards of scholarship, and until recently it had been largely assumed that anyone who met those standards was naturally able to teach well. That myth had broken down, and now the hires had to truly be able to do both. Of course, leeway was granted on the teaching part, but hiring excellence in research never budges. My first administrative energies were directed toward that goal.

Hiring at universities is a tediously complex process requiring input from the faculty, and I mean all of the faculty. They make the special interest groups in Washington, D.C., look like warm-and-fuzzy, bipartisan zealots. In the world of academics, everybody has a special interest. How was I going to consolidate all the special interests into a coherent program? In what area of neuroscience would we focus? And which new hires would be based in what academic unit? Suddenly, the view from the ground made it all look impossible.

Leo knew all of this, of course, because he had been the center's architect. As a veteran member of the campus, everybody knew him, and he knew everybody. Academics can be bland, passive, detached, aggressive, obstructionist, you name it. Not Leo. He is a vibrant personality who engages the mind. Sometimes the message wasn't liked, but everybody liked him, and they trusted him. Leo was working the faculty, bringing them along all the time. He had placed his bet on me, and he was going to help make it work.

A plan was struck that pleased everybody. They basically said to me, "Mike, start bringing in people, and we will offer jobs to the best. If you screw up, we will let you know." Trust once again prevailed, even though we still hadn't chosen an area of concentration. The idea was to get quality, and the rest would take care of itself. I took my hunting license and went to work.

All of us kept talking about the pool of ten jobs we had, but we tended to forget that each of those jobs was pegged to a particular academic unit. At first I maintained that where people would be academically housed was a mere detail. I have discovered in life that I can be wrong, then dead wrong, and then so wrong it is hilarious. In this instance, a new candidate would come into town, be interviewed, and give a job talk. Then, with quick polling in the hallways, I'd get the consensus from my new colleagues as to whether the candidate was desirable from their point of view. Depending on that quick calculation, I would then do the soft sell up in my office. Not surprisingly, the candidate would ask, "What department would I be in?" I would say, "Well you have a choice. You can be either in the Medical School, the Biology Department, or the Psychology Department." They would then ask, "What's the difference?" I would say, "In the Medical School it is an eleven-month appointment, it is more money, and you basically don't have to teach. Now, in the Biology Department it is a nine-month appointment and you have to teach one course. The Psychology Department is also a nine-month appointment and you have to teach three courses." After the slightest of pauses, most candidates would then say, "Which Medical School department would be best for me?" I quickly learned to offer no choice and tell them up front which department they were being recruited for; that sped the hiring conversations along with few or no problems.

While there was implicit pressure to do something big, to make a splash in the scientific community by hiring a well-established person, I ultimately pulled back from this approach. It turns out that 85 percent of what are called "senior hires" never work out. The process, in terms of the time invested and of the inordinate amount of wooing (from weeks to months to years), is expensive and usually ends up nowhere. A program can be held hostage to a big name, and in the end, nobody has been hired and the program trying to be built is injuriously delayed. This was all explained to me by those more knowledgeable, but my outrageous optimism found me trying anyway—but just once. I now believe those statistics! The big shots don't usually have to leave where they are, and when push comes to shove, they suddenly adopt a wish list that would not be granted to the queen of England. If the wish list is granted, they take the job, and if not, they take the easy route by staying put, usually scoring a handsome retention package. The most painful part is listening to the post-decision rationalizations as to why they "feel" they couldn't accept (the interpreter talking).

My solution was simple enough. The senior stars of the field are stars because they train really talented people. If a particular lab has been putting out particularly interesting work, there is a good chance that the lead graduate student or seasoned postdoctoral fellow in that lab is the person you want. They are young, less expensive to establish, full of zip, and usually looking for a job. In addition, a job in the UC system was coveted. So, one contacts the successful scientists to see who is on deck for a job and makes sure they know about our openings. It worked like a charm. It wasn't long before a stream of unbelievably talented and eager young scientists started to march

FIGURE 36. The fabulous eight young professors first hired at the Center for Neuroscience, UC Davis: Barbara Chapman, Charles Gray, Ken Britton, Leah Krubitzer, Bruno Olshausen, Greg Recanzone, Ron Mangun, and Mitch Sutter.

though the center. It was easy to pick the best of the best. They pop out from the background in some hard-to-define way. The decisions on whom to hire were always unanimous (Figure 36).

THE PATIENT'S TURN TO TRAVEL

It wasn't long before I knew we would continue to study split-brain patients. J.W. and his wife seemed open to the idea of moving to Davis, too, even though California was a long way from their tiny hometown in New Hampshire. J.W. knew California as a result of his trips to La Jolla, and he had always liked it. With the number of fellows and graduate students now working for me at UC Davis, we could keep J.W. busy every day. Through the administrative genius of my project director, we were able to compensate J.W. and his wife enough to live in a very nice house in South Davis (Figure 37). It worked for a year and a half, until they became homesick. As it turned out, he was a little over two years ahead of my own yearning for New England.

FIGURE 37. J.W. at home in Davis with his wife. J.W. loved to assemble model cars, and he has sustained this hobby all his life. We picked them up in New Hampshire, moved them to California, and moved them back again.

Being able to test J.W. almost daily led to dozens upon dozens of experiments. One of the first young scientists hired was Ron Mangun, who had been my postdoctoral fellow at Dartmouth. As the new director, I was given one appointment out of the ten as a courtesy, but also because those administrators knew I would be busy dealing with stuff that I hadn't fully anticipated. I immediately asked Ron, large in body, mind, and heart, if he wanted to come back to California and start a new life. He jumped at it. I had my first hint that everybody wants to move to California, especially if they are young and ambitious (Figure 38). Soon enough, Ron would be hired away from our UC center by Duke University to run their new center. Later, however, he returned to Davis to run its new Mind/Brain Center, the sequel to the Center for Neuroscience. Now he is dean of social sciences at Davis.

As an expert in the field of electrical recordings of the brain, Ron began to study whether the flow of neural activity could be tracked by measuring when certain parts of the brain reacted to a light presented to only one visual field—a simple experiment. We knew that in the normal brain, information had to be communicated from one side of the brain to the other. We also had the continuing debate about possible subcortical communication between the hemispheres.

Maybe recording actual neurophysiological signals could illuminate the question. It was already known that certain brain waves accompanied the processing of visual information, something called the P1/N1 complex.[1] These waves were easily detectable and amazingly symmetrical over both visual hemispheres, even when a stimulus was presented to only one visual field. How and through what nerve circuits was that all getting done?

Remember, if a word is flashed to the left visual field, the information travels directly to the right hemisphere. If you've named it out loud, that information has already flowed from your right visual cortex over to the left brain speech hemisphere. Could that flow of electrical activity actually be detected? Simply doing the scalp recordings was the easy part. Anybody who has had an EEG knows the drill. A little squirt of conducting jelly is placed on the skull, and a sensitive electrode is mushed into it. It is connected to a preamplifier/amplifier and ultimately to a computer for analysis. That is the easy part.

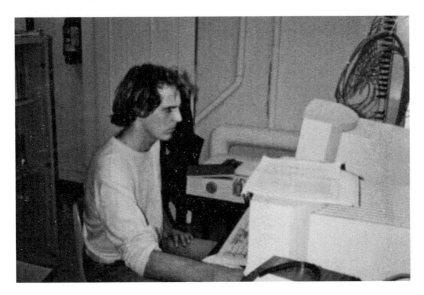

FIGURE 38. Life for young academics is uncertain, as talent and mentorship are hard to find. Here neophyte Todd Handy, who originally wandered into our labs in Hanover, works in the basement of Dartmouth's Pike House. Soon he developed a passion for research and followed Mangun to UC Davis for his graduate work. When I left Davis for Dartmouth, I convinced Handy to join the new, swanky labs Dartmouth had built for us. He is now a professor at the University of British Columbia.

As expected, Ron showed that the signs of the signal first appeared in the right hemisphere. Then, a few milliseconds later, it appeared in the opposite hemisphere as well. It was clear as a bell. He then found a patient who was going to have full callosal surgery, but in stages starting with the anterior areas first. When he did the experiment, the P1/N1 complex continued to flow from the right hemisphere to the left hemisphere after the first surgery. In that patient, it took three surgeries to get the entire callosum sectioned. After the second surgery, which moved the disconnection ever farther back toward the posterior regions, there was no effect on the brain wave. Because some fibers had been inadvertently spared, there was a third surgery, which completed the sectioning of the callosum, and the brain wave no longer appeared in the left hemisphere, only in the right hemisphere, which first received the stimulus. That cinched it. Now there was no question: The synchrony of brain waves in our cortex was due not only to an identifiable brain structure—the callosum—but to a specific region of that brain structure. It was also clear from his data that one could track the timing of the exchange of the information.[2] All of this work was confirming something else in our minds: With enough cleverness, both the place and the timing of the underlying neural processes guiding a psychological phenomenon, such as visual attention, were at hand.

While J.W. participated in Ron's experiment, he was also involved in several others. The sheer convenience of having him, with his incredibly cooperative nature, live around the corner made him the dream patient to study. He was smart, humorous, and respected by all the researchers. He could dazzle at one moment with an easily visible skill that seemed only a split-brain patient could do, while at the next moment be part of a study that required hundreds of trials to extract some kind of significant response difference in reaction times. Those kinds of studies revealed many interesting phenomena, but they were only evident in the data analysis, not as some kind of obvious and unique behavioral skill.

One such obvious skill was the ability to do two things at once.

Jim Eliassen, a graduate student who had come with us from Dartmouth, was captivated with the idea that J.W. might be able to guide each hand to do conflicting tasks. Eliassen, a clever and canny Stanford graduate, was unfailingly a gentleman. He always brought a level of good cheer, mixed with a sharp and unwavering eye for experimental detail. While he meticulously quantified everything, the phenomenon Jim studied was so behaviorally evident that it made it all the way to a PBS special, hosted by Alan Alda. They set up the study by showing Alda trying to do Jim's task.

Imagine yourself, armed with a pad of paper and a pencil in each hand, sitting looking at a dot right smack in the middle of a TV screen. Simple visual geometric shapes will be flashed, and all you have to do is draw them simultaneously. Easy, right? It is easy only if both of the pictures flashed are the same thing. So if two circles are flashed, no problem. If a circle and square are flashed, however, it's a big problem for you and me and Alan Alda. We start, then stop, then draw something out of whack. None of it is done simultaneously: Instead each hand works in an alternative style. In short, with this little task, the human and his great big brain seem totally flummoxed.

Now J.W. is asked to do the same thing. The two circles, no problem. The square and the circle: again, no problem and done instantly (Video 14). It was as if two people were present, one guiding each hand with absolutely no interference from the other. No statistics are needed to see this effect, though thorough experimental description is always required to pick apart the spatial and temporal aspects of the seemingly unfathomable skill, and this was done as well.[3]

ANOTHER RIGHT HEMISPHERE SPEAKS UP

While literally dozens of experiments were going on, most of them showing how poorly the two separated half brains intercommunicated both sensory and motor information, another development was also occurring right before our eyes. It looked like J.W.'s right hemisphere was beginning to speak. Although it had been quiet for years, like a late-developing child, suddenly one-word utterances were coming forth. When this happens, it begins to feel like the person is no longer split. Each side describes and reacts to its own sensory sphere, and together they look whole to the outside world. After all, within a half brain hundreds to thousands of modules interact to produce that half brain's mind. Maybe mind left and mind right, though separate, look unified not only to the outside observer but also to the inside observer.

Kathy Baynes decided to figure out what was going on with J.W. She knew all the patients, having started with us in New York and traveled with us to Dartmouth and now to Davis. She is one of the people who truly make universities and research centers work. She is collaborative to the core and generous with her huge intellectual skills to projects that are not even central to her passion, the study and nature of human language. The bait on the hook for her was language, of course, and J.W. seemed to be undergoing some language changes. Was this brain plasticity at work? Was it a clever way to cross-cue? Was sensory information transferring over to the smarty-pants left hemisphere? The first thing to do was to carefully define J.W.'s changing behavior.

As you know by now, one of the cardinal tests for splitness is to flash pictures of objects, words, anything to the left and right visual fields. If a patient is speaking only out of the left hemisphere, only stimuli in the right visual field should be named. If the patient starts naming left visual field stimuli that are only streaming to the nonverbal right hemisphere, something is going on. What was it?

J.W. had undergone his surgery at twenty-six. By now his left hemisphere had been separated from his right hemisphere for fifteen years. For most of that time, the left hemisphere had been doing all the talking. If asked questions, it answered them while the seemingly patient right hemisphere came along for the ride. When J.W.'s right hemisphere spoke up, we were all stunned. Was the right hemisphere going to be like Sleeping Beauty and dazzle us with stories of forbearance? Was it going to show differences in intellect, capacity, personality, and indeed, be a different kind of agent?

In the last chapter, I mentioned a BBC special on conjoined twins,[4] which is a remarkable story of love, normality, and gumption under the most unlikely of circumstances. The twins, Abby and Brittany, are conjoined at the chest and torso, and have a single pair of arms and legs. Although Abby controls one arm and leg and Brittany the other arm and leg, they are athletically coordinated. Scientifically, they are two people locked into one body. They each have different desires, different likes and dislikes, and different personalities. It is as odd as it sounds, and yet their life seems natural to them. The circumstances are their normal circumstances: The person adjusts, normalizes. The trick is to get others to allow them to be normal. This is where their parents (who have raised them brilliantly) and their friends excelled. Brittany and Abby have conversations with each other all day long. They have their differences, but they are also incredibly able to cooperate and cue the other on matters that will assist in the management of their commonly shared body. In short, there is independence yet cooperation.

Now imagine yourself standing directly behind another person and then being tightly taped together. It is a good taping job, such that when one head moves left, the other is perfectly yoked to it and winds up looking to the left as well. When a hand moves, the taped hand of the partner moves along with it. Again, there is no question there are two separate and equal minds guiding what amounts to one body. How long would it take to learn to talk to the other, so goals are shared? How long would it take for strategies to be formed so that the motor commands to one hand from its brain would serve as a cue to the yoked hand? For example, when left hand A wants to move to the left, left hand B somehow needs to be cued that it should relax and go with the flow: Hand A would start to move, and hand B would sense this through proprioception* and learn to go with it. How long would all of this take? In J.W.'s case, approximately fifteen years.

With J.W., it didn't happen all at once. One could say it took practice, practice, practice. Approximately seven years after his surgery, J.W. was able to do a curious thing. His left hemisphere speech apparatus could name which of two numeric stimuli had been presented to his right hemisphere. Oddly, his left hemisphere didn't know and couldn't access the information for internal use. Let's say I had flashed the number 2 to his right brain. I then flashed the number 2 directly to the left hemisphere. Could the left hemisphere say "same" or "different"? It could not, and on such tests the left hemisphere performed at chance. Yet, if I had asked the left hemisphere to simply speak, it would respond correctly and say "2."[5] Weird!

Somehow the right hemisphere was setting up the speech apparatus, but in a special way. At this point, both hemispheres had to know what the possible answers were in each test. Additionally in this experiment, there could be only two possibilities, "1" or "2." Again, somehow the right hemisphere could set the speech apparatus to one of two possible reactions. It seemed it must do it down low in the overall speech apparatus mechanism, since the left brain cognitive mechanism seemed ignorant of the information. We ran a further test to show that the only thing at play was the capacity to set the speech system with one or two possibilities for response. As long as the two possibilities were known, this interhemispheric parlor trick worked. Instead of asking J.W. to say "1" or "2," I then asked him to say "indescribable" or "indestructible." Again, both hemispheres knew what the two choices were. Asked to compare the words as same or different, both hemispheres failed, but when the

---

* Proprioception is the awareness of the position of the various parts of the body. It arises from stimuli from sensory nerves in muscles, tendons, and joints.

word was flashed to the right brain, the left brain correctly managed the spoken response.

After another seven years, J.W. had begun upping his game. Now he was naming about 25 percent of the pictures flashed to his supposedly mute right hemisphere. We started out using pictures of familiar family and friends, as well as pictures taken from a standard set of pictures of neutral objects and animals. We presented the stimuli in both the quick-flash method and in our eye tracker, which allowed for the picture to remain in full view for up to five seconds. The results were clear. J.W. was naming the stimuli about 67 percent of the time and it didn't matter if they were flashed or remained on for viewing. The right hemisphere had definitely changed. Its control of the speech apparatus was now good (Figure 39).

Even more surprising was J.W.'s seeming ability to describe what we called "complex scenes" flashed to his right hemisphere. While he never could name the scene completely accurately, he did pick individual elements in a picture. For example, he initially identified one scene correctly, but then incorrectly described it after being prompted for further information. The scene depicted a woman wearing a black dress and washing clothes in an old-fashioned washer. Behind her was a clothesline with laundry hanging from it. Here is what J.W. said (the experimenter's comments are in parentheses):

FIGURE 39. Here we tested J.W.'s capacity to develop speech in his right hemisphere, by presenting him with images to name. To ensure he wasn't cheating, we used an image-stabilizing eye tracker. Here he is viewing the display screen through the stabilizing device with one eye. His other eye is covered with a patch and his head and chin are secured by a bite board for maximum control.

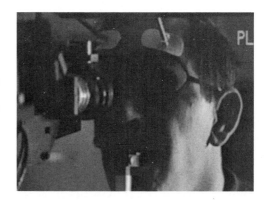

It was a person . . . Would it be someone hanging out their laundry? One person. Must have been a woman. (Did you see any laundry?) I think so. I think she was reaching up and that's what she was doing. . . .

Take another example, which was a scene depicting a woman standing behind another woman who was sitting at a table and crying. There was a stove and sink in the background. In response to this scene, J.W. said:

The first thing I thought of was a woman baking. I don't know why. . . . (Was she sitting or standing or . . .) Standing up by a table or something.

J.W.'s response does not capture the meaning of the scene, but it does convey some of its visual attributes.

Another example shows how J.W. could capture both visually and semantically similar information about the stimulus. The scene depicted a racetrack with two racing cars driving around it and another car that had crashed and overturned. There was a grandstand off to the left behind the track.

Looked like something moving like a vehicle or something or somebody running or something like that. (Did it look like one thing or . . .) At least one. It was centered on one. Maybe there was something in the background. (If you had to guess what it was, what would you guess?) Either somebody running, or, a curved picture. Looked like coming around a corner almost . . . someone running. Maybe it was a track. It was hard to tell.

The level of performance looks like a possible response from a less capable mental system. After all, language is not what the right hemisphere normally does. Is this a fumbling infantlike language system in a grown-up body/brain? It did appear that J.W.'s naming performance became worse as the visual stimuli became more complex. J.W.'s performance level was indeed like other patients who had also developed speech after their callosal section. In all cases, utterances from the right brain seemed to be only single-word responses. Producing multi-word descriptions was seemingly not possible for J.W. Yet it sounded like he was making them.

We came to see that this kind of capacity has a collaborative strategy between the two hemispheres. We knew that there was no interhemispheric transfer of information mediated by pathways in the brain. We also knew J.W. had no syntactic ability in his right hemisphere. While his language was extensive, his right hemisphere couldn't tell the difference between "a blind Venetian" and "a venetian blind." Word order had no impact on him, as Kathy had determined in test after test.[6]

So how was he doing it? We finally figured out that the collaboration could consist of the left hemisphere generating complex descriptions based on one- or two-word "clues" generated by the right hemisphere.[7] It's like the old couple again. One is droning on, and the other pops in with a single word to get the narrative back on track. The droner notes it, continues with the pitch, makes adjustments, pauses, only to get another single word from the partner, and on and on.

After intense study in Davis, J.W. unfolded in slow motion, making it easier to see his progress. When speech developed in P.S. and V.P., it happened quickly, within one year, and it seemed different. J.W.'s progression was slower and also reminiscent of the original Caltech series of patients.

LEFT BRAIN/RIGHT BRAIN DIFFERENCES FROM CELLS TO PROCESS

One of the great values of working in an interdisciplinary center is that experiments and questions come to mind that otherwise might not. Leo had trained a young neuroanatomist, Jeffrey Hutsler, and I had the good fortune of being able to hire him. He was exceptionally bright, a tinkerer who got things to work in the lab and, paradoxically, was hilariously funny yet also a bit of a loner. Getting things to work in a cellular and chemical neuroanatomy lab is a lonely job. It takes lots of time and patience and sheer hard work. It was only because Jeff and I happened to be in the same place at the same time that we were able to go after one of the underlying questions of human brain function: Is there something special about the human cortical areas that enables language?

The question of "nature versus nurture" pops up often in the field of brain science. Do things come wired up with little possibility of change? Or is a framework laid down by genetics but modifiable? In the lab next to Jeff's, we were watching with our own eyes the stirrings of language and speech from J.W.'s formerly mute right hemisphere. Had the underlying wiring changed or had it already been there and begun reemerging in a way we didn't yet understand?

There was another part to the puzzle. The split-brain patients had all been carefully studied by Hillyard and his colleague Marta Kutas. Marta was soon to become the world's expert on a particular brain wave she had unearthed, called the N400, fondly referred to as the "semantic incongruity" wave.[8] Participants in her tests would listen to a sentence like "I take my coffee with cream and . . ." and then an ending would be flashed that was either congruous, "sugar," or incongruous, "cement." When this was done on normal subjects, the brain answered back with the N400 brain wave response when the word was incongruous, but not when it was congruous. Marta had in essence found a brain biomarker for syntax.

People who study electrical brain waves also want to know where the waves are coming from. There is a long and complicated story about the "generators" of brain waves. In normal subjects, the N400 wave was present over both cerebral hemispheres, even though only the left hemisphere was believed to be involved in language comprehension. What would happen with split-brain patients? Kutas and

Hillyard studied five patients overall, and the result generated even more intrigue.

In many previous studies, the five split-brain patients studied had all shown evidence of language in their right hemispheres. Thus it was of little surprise that each half brain could detect a semantic anomaly and could indicate it by pointing with the appropriate hand to the word that made no sense. What was so surprising was that only two of those same five patients produced an N400 wave when an incongruous word was shown to the right hemisphere. At the time of these tests, these were the two split-brain patients who had shown evidence of right hemisphere speech.[9] Had Marta found some kind of marker associated with only a fully realizable language and speech complex? Were the neural processes that enabled this brain wave already there or had the right hemisphere learned to speak? Maybe language alone didn't do the trick. Maybe the wave appeared only when the hemisphere could comprehend that, indeed, the incongruity was semantic!

It was time to go under the hood. Could Hutsler find out some basics on normal postmortem brains? Is the language cortex special and if so, where is it? We started by simply comparing the cortical areas in the dominant left hemisphere that were classically defined in the neurology books as being associated with speech with the corresponding area in the right hemisphere. Of course, to do this, we needed to study postmortem brains. Enter two dynamos from the Veterans Administration hospital in Martinez, just south of Davis: Robert Knight and Robert Rafal. By objective standards, these were two of the best behavioral neurologists in the country. They were also a lot of fun. Unlike many of my colleagues, they also loved to teach.

Karl Lashley was not quite on the mark when he told Roger Sperry that if you have to teach, teach neuroanatomy because it never changes. Lashley was right that the major lobes, fiber tracts, and major nuclei are normally situated where they are supposed to be. That is, the basic infrastructure of the busy brain city is all there:

all the stores, roads, and alleys. Where he got it wrong was in the details of the connections. Neurons do change and vary where they do their business . . . all over the place. But first, Hutsler wanted to know if the cellular organization of the left language cortex was different from the cortex found in the normally language-absent right hemisphere. He found many differences in the underlying cortical organization of the left and right hemispheres that, overall, suggested the cortical neurons were making many more connections to other neurons on the left side. He did not find there were more neurons, per se.

BUILDING AN INTELLECTUAL COMMUNITY

The UC Davis community was hopping with activity. One way we enlivened the tomato patch was to bring in visitors. Our new center building was up, and I had adorned the walls with the paintings by Henry Isaacs, a gifted landscape artist from Vermont and a dear friend. The new hires added their own huge energy, and the Davis faculty supported the enterprise completely. Still, it never hurts to add dessert. I cooked up an idea to have a visiting-professor program.

The first requirement of a visiting professor is to invite a truly outstanding intellect: someone who has the knack for real interaction and who works and thinks at a very high level. Someone with such capacities is welcomed by all factions because everyone realizes they will learn something. I went after Endel Tulving, the distinguished Canadian expert on human memory. I had heard he might consider spending some time out of Toronto because, even though he was retired, the University of Toronto had mishandled his pension and had left him shortchanged. I had met him only once before, at a New York meeting, and we immediately took a shine to each other. I was soon to learn he was a native Estonian. He had escaped from the horrors of the Red Army when he was seventeen and migrated

FIGURE 40. Enjoying a moment with Endel (front left) and Ruth Tulving (front right), their daughter Linda (standing), and other lifelong friends. The Tulvings were passionate supporters of Estonia and bought an apartment there after the breakup of the Soviet Union.

through Germany, where he finished his high school education and then moved to Canada. It was in Germany that he decided he wanted to be a psychologist.

Still, we needed enticements to lure him to Davis. A short walk from our now-beloved center was an apartment complex. I asked Buck Marcussen, my administrative assistant, who could get an elephant to fly if need be, how we could rent one, given that UC needs four thousand approvals for everything. He figured this out. Then I went to Leo to ask how we could get an appointment for a visiting professor on an annual basis. He knew a way. Finally, I turned to one of our young fellows, the captivating and articulate Helmi Lutsep, a young neurologist studying stroke patients, and an Estonian! I knew she would make Tulving feel welcome. It even came to light that Helmi's father had gone to high school with Endel, meeting in a displaced persons camp in Germany after World War II. Helmi got hold of an Estonian flag and hung it up next to a picture of Endel as a young man, which had made its way from her father. I don't know what he expected Davis to be like, but certainly it surprised him. We all became fast friends and have remained so for twenty-five years (Figure 40).

Tulving's contributions to the field of cognitive science were already legendary. He had firmly established that we humans enjoy at least two kinds of memory, semantic and episodic.[10] Semantic memory deals with things we learn, such as the rules of the chess game. Episodic memory occurs when you remember actually playing a particular game of chess: the experience and the episode, so to speak. Tulving was now going after the question of whether these two kinds of memories were located in different brain regions. The new brain imaging technologies now existed to test such questions, and he tackled them with the energy of a twelve-year-old. His door was open at 9 A.M. and didn't close until 7 P.M. His intellect, energy, and, best of all, character were always on display. And right behind him was his adoring and patient wife, Ruth, who was a world-famous artist in her own right and a member of the Royal Canadian Academy of Arts.

Ruth became rather famous locally for her paintings, which she called "Butterflies of Davis." They were intricate, beautiful, and stunningly original. They became the signature for their time in Davis and now hang all over the world. What is so interesting about Ruth and Endel is that while she seemed almost invisible with Endel around, she was the engine that kept it all going, from gracious hosting, to fabulous meals, to supportive conversation, to just about everything. Endel would lovingly say, "My bride has put up with me for fifty years."

Ruth died in 2013 after a two- to three-year illness that robbed her of her memory, her ability to speak, and ultimately her ability to cogitate. After enjoying a lifetime of shared intellect and interests, her condition was especially heartbreaking for Endel. Endel's response was powerful. Up until her illness, he'd never lost a minute working on his passion, the problem of human memory. When it became apparent Ruth was truly ill, when all the clever dodges patients use to hide their internal deterioration were revealed, Endel stopped all of his work and cared for her in a way that was both moving and seamless. Not a moment of pity, regret, complaint, or despair. As Endel put it to me, "Ruth took care of me unselfishly for fifty years. Now it's my time to take care of her. Let's have a Manhattan."

Endel was a magnet to students. During one of his terms, he discovered a new student of mine, Michael Miller. Mike wanted to study false memories, yet in good graduate school fashion, he was mandated to take rotations outside of his area of interest. He somehow managed to be reassigned to Endel, and as Mike puts it, "I certainly learned everything I know about memory from him. I used to sit in his office for hours while he talked about episodic memory. We even devised a little paper-and-pencil memory test on orthographic distinctiveness* that I tested on a bunch of subjects. It was a blast."[11]

What happens in full view when expert meets motivated novice is so exciting and fulfilling to watch. The novice can't get enough of the expert, and the expert needs a jolt from the novice. Usually, both people, if they are good-spirited, learn.

When Endel began his Davis visits in 1993, he was totally consumed by the new brain imaging data, especially that coming from PET studies, which measure the parts of the brain that are metabolically active when doing work. Endel was interested in the moments when the brain was either encoding new information or retrieving stored information. Add to that Endel's famous distinction between memory for events (episodic) and memory for knowledge (semantic) and you have a complex question on the table. Do these processes call upon the same brain systems to work or will the fancy brain-imaging techniques reveal that the processes are managed by different parts of the brain? Spending a little time on this leads to many, many complicated questions.

---

* Orthographic distinctiveness is the structural characteristics of a word that make it physically unusual, interesting, or distinctive.

Take polysemous words, a single word that can have several different meanings. My favorite is the word *line*. One can get in a "line," read a good "line," have a "line" on a great deal, walk the "line" with Johnny Cash, and so forth up to twenty-six other meanings. Some of those meanings are more likely used for episodic memories and others for semantic memories. Are the twenty-six meanings of the word filed away separately in the brain or is there one storage spot and other mechanisms that come in to give the word context?

Psychologists are among the world's most clever experimentalists. While the physical and biological scientists are dealing with stuff that is tangible, psychologists are dealing with "abstractions," or "representations," or "attitudes," and on and on. It is a much more difficult science, much more slippery, and the experimental designs to get at it are much more trying. Endel, with youthful enthusiasm, was wading in to the PET measurements with a sense that he would be able to provide new and hard evidence that indeed multiple systems in the brain carry out these various yet seemingly similar tasks.[12] His impeccable standards of science and his persistent energy in relation to a question sucked in all of those around him. Naturally, Mike became intrigued. Even though he was working on other problems at the time, he decided to work on this one, too. Over the next few years an army of people jumped into the discussion.

Solely using the imaging data, Endel and collaborators, who by now were strung across the country, originally unearthed a pattern of brain activation that suggested that the left and right prefrontal lobe regions of the human brain were separately involved in the encoding and retrieval of episodic and semantic memory.[13] The prefrontal areas, those territories way at the front of the brain that were not commonly thought to be part of memory circuits, were, they discovered, active during memory tasks. That, in and of itself, annoyed many people. But there was more. Endel's team felt that the left prefrontal areas were more active when a subject was retrieving verbal semantic memories, while the right hemisphere prefrontal regions were more active when retrieving episodic memories. Added to this was the idea that the left prefrontal areas also kicked in when encoding episodic information of a novel kind. Whew. Distinctions, distinctions, and more distinctions.

One can immediately see the complexity of possibilities here, and the field of cognitive neuroscientists who worry about this kind of thing are still hashing it out over the paper written while Endel was at Davis.[14] I was in the background basically watching all of this when Mike suggested we get off our duffs and test the idea on split-brain patients. Up to that time, the issue of memory processes in these patients had languished with only a couple of observations. Not surprisingly, these had been two conflicting observations. One study suggested that the patients experienced memory deficits[15] and another that they did not.[16]

The question came down to whether the two half brains behaved differently now that they were disconnected from each other. The Tulving model would suggest that the left hemisphere would not be as good at retrieving episodic memories and that the right would be poor at retrieving semantic memories. What Mike found was something different, and this is where split-brain patients offer such elegant answers to complex questions.[17] One can simply ask a disconnected left or right hemisphere what it can or cannot do. In this case, Mike showed that each side could pretty much both encode and retrieve information of all kinds. What he also found, however, was that the left side was better at verbal information, and the right side at handling visual information, such as faces. In other words, each hemisphere has its specialization. The hemisphere that specializes in the kind of information being presented performs better on that kind of information. It has nothing to do with semantic memory versus episodic memory.

Overall, it was beginning to look like each hemisphere's recognition memory* system behaved similarly, as long as the testing stimuli

---

* In contrast to free recall, recognition memory is the ability to recognize previously encountered objects, events, and so forth.

were not playing to a particular hemisphere's strong suit. Each side could encode and retrieve new information. Having said this, however, I am left wondering. Endel is rarely incorrect. I should probably plan on doing more experiments.

TRAINING THE NEXT GENERATION

Everywhere I went for almost thirty years, people were interested in the split-brain patients and what they were teaching us about brain organization. Anatomists; physiologists; developmental, perceptual, and cognitive scientists; evolutionary psychologists; and philosophers all wanted to know how their work might be illuminated by the studies. The added benefit of all this attention was that I came to know people across a broad range of scientific fields. The vast fields of neuroscience and cognitive science rarely mingled, and the leaders of all the specific disciplines hardly knew each other. Yet I found myself knowing them all. And I knew as plain as day that they would all benefit from working together.

I had begun managing the summer institute in cognitive neuroscience at Dartmouth Medical School in 1989. It was originally funded by the James S. McDonnell Foundation to bring together on a regular basis investigators from disparate fields and young scientists being trained in these fields. The notion was to expose them to a broad base of methods, approaches, and bodies of knowledge from neuroscience, medicine, psychology, linguistics, computer science, engineering, and philosophy. When we moved to Davis in 1992, we had arranged that the summer institute move as well. Leo Chalupa and Emilio Bizzi took charge of the two-week school, which was now in its fourth year of existence. The idea was also born that we should meet every five years to start a regular stock-taking of the field by specialty. To each meeting, each specialty area would invite eight additional scientists, all leaders in their discipline, to have an intensive meeting of self-examination. That calculated out to almost one hundred scientists coming to one place for three weeks. I thought it would be grueling, if not impossible. We wanted each participant to write a chapter about their work, and we had arranged that MIT Press would publish it all in one huge reference book.

With all due respect to Davis, it seemed unlikely that its 110-degree summer was going to be alluring. We were worrying about where we might pull this off when in walked a young Brit, the vibrant Flo Batt. She was the original let-me-solve-it person with a lovely British accent to boot. We hired her to manage the complex project. After just a few weeks, I was sure that she must have worked for the queen of England. She wrangled and sweet-talked and cajoled. Everything worked out for us to have the entire three-week event at the Resort at Squaw Creek, near Lake Tahoe in the Sierra Nevada mountains, which was the right place to be in the summer of 1993 (Figure 41). It was and is fantastic. We continue to have a meeting there every five years, with all the goodwill that was set up by Flo. The most amazing thing about the experience is that I sit at the back of the room for every lecture. That is not my usual way, but for some reason, this event is like no other.

FIGURE 41. The very first Tahoe meeting lasted for three weeks. Virtually every fellow here has gone on to have a distinguished career in the field.

The book that comes out of this meeting every five years has become the reference standard in the field. Its consumers, the working cognitive neuroscientists, have even been caught reading chapters outside their own fields. The idea worked, and it worked because the section editors and the publishers were all behind the effort. Starting with the second edition, consisting of all new chapters and several changes of participants, my daughter Marin became the managing editor in her spare time.[18] Meanwhile, my wife was running the *Journal of Cognitive Neuroscience,* and my daughter Kate was launching the new Cognitive Neuroscience Society. It was beginning to look like a family business.

Deciding to launch a scientific society is a very iffy business, full of doubt, politics, money concerns, and just plain angst. What if nobody shows up? Who is going to underwrite the set costs of hotels, coffee, and meeting rooms? What if the leaders in the field don't show up? What if it is a bridge too far? We talked it around and around. Finally, it was time to act.

My old formula was put in place. Let's pick a place people like to visit. San Francisco it was. Let's look at a classy hotel, so I went to check out the Fairmont. They had a room that would hold four hundred, which at the time looked like the Los Angeles Coliseum to my worried mind. I began to learn about hotel contracts, minimums for room use, and on and on. They needed a deposit, so out came my Visa card and, gulp, five thousand dollars was plunked on the table. The Buckley/Allen debate at the Hollywood Palladium flashed through my mind, but without Steve Allen to slake my anxiety.

The die was cast. Now it was time to nose around for some exposure. Just like in the movie *The Magnificent Seven,* calls went out to friends across the country. Mike Posner at Oregon (who was awarded the National Medal of Science in 2009) joined the effort, as did Dan Schacter, who was then at Arizona and about to become one of the world's experts on human memory; Patti Reuter-Lorenz, my postdoctoral fellow, who is now a professor at the University of Michigan; Art Shimamura, then an assistant professor

FIGURE 42.
The very first
Cognitive
Neuroscience
Society
meeting in San
Francisco's
Fairmont Hotel.
The keynote
speaker was
Steven Pinker.

and now a full professor at Berkeley, a memory expert who is also interested in art and aesthetics; and, of course, Ron Mangun at Davis. I asked Steve Pinker, now one of America's most prominent psychologists and public intellectuals, if he would be the keynote speaker (Figure 42). No problem. I then talked up the meeting, as did all my Davis colleagues, and, in the end, an absolutely stellar group of scientists showed up for talks and separately scheduled poster-style exhibits showing their research (Figure 43). Critically, key members of my staff from Davis caught the fever and came in to help manage the meeting. As it turned out, there had been little to worry about, as timing is everything, and everybody wanted the idea of cognitive neuroscience to work.

More than four hundred scientists showed up and the excitement was palpable. My idea for the society was that the administrative processes would be invisible to the participants. Let the scientists come and thrive on the science. Let's not have business meetings and committees formed to worry about this and that. I guess it

FIGURE 43. The number of distinguished scientists who showed up at the meeting to support our efforts was stunning. Here, with the mustache, is V. S. Ramachandran, speaking between the shoulders of Richard Andersen (*center*) and Michael Merzenich (*right*).

was naïve to dream of this scientific nirvana. Within a day, there was a spontaneous movement to hold a "business meeting," and dozens of people showed up demanding representation and process and, of course, gender balance. Since it was pretty much my daughter and I who were running the society, we had nothing going for us but gender balance.

Over the years, the Cognitive Neuroscience Society has grown and thrived. First, the appearance and then the reality of structure took hold and with it more ideas for speakers, themes, diversity, and ultimately intellectual specialization. The original idea of the meeting was to have all participants hear a set of common lectures, the idea being that it would be broadening for scientists interested in topic A to hear and keep up with advances in Topic B. Both the content and metaphor of other specializations can be enlightening. That simple idea has been almost impossible to realize. People have a deep need to promote their own knowledge and tend to have mental space only for the specifics of their own field. Interdisciplinary work is talked about all the time but is rarely achieved. Twenty years on, the society experiences the full grip of special interests and, unavoidably, becomes a mosaic. That is the way it goes. I have learned that I am an entrepreneur, but not a particularly good manager. I need to build it, and then gradually turn it over to others to run, even though it is emotionally draining to do so.

Davis was fun and thriving. Why would anybody think about moving? The new hires were all in place, filling the Davis community with intellect, ideas, and work. Close friends had been made and the family was happy. We had lunches in Napa, weekends at our cabin in Tahoe, and visits to San Francisco. We were knee-deep in funding, weekly lunches with Leo at Biba's, Sacramento's only great restaurant, and regular visits from Endel and many others. Most important, we had a happy family. Yet through it all, I felt an irrational yearning for Dartmouth. It had started as an undergraduate, disappeared for twenty years, then popped up again when I worked at the medical school.

This time, the college called me. Dartmouth, upon my departure in 1992, had invested in their cognitive neuroscience psychology department by hiring the distinguished neuropsychologist Alfonso Caramazza. He had arrived in town in his dazzling way, demanding better science and better food! As far as I knew, he loved it there and became the face of psychology to the administration. Lee Bollinger, the provost at the time and now president of Columbia University, took an immediate shine to him, and he instantly became part of the landscape. It looked like all was well.

For reasons as inexplicable as my own, Alfonso left for Harvard in 1995. He was a hot property, and his new wife was a lawyer who had gotten the itch to live in Boston or something like that. For whatever reasons, Dartmouth's cognitive neuroscience slot opened up, and they started seeking a replacement. They first invited Harvard's distinguished psychologist Daniel Schacter (who had moved there from Arizona) to visit. A little tit-for-tat must have been in play. Dan was happy where he was, so he left them with the message that they should try to get me to return. Good friends at the college promoted the idea, and, while always open to any ideas about Dartmouth, at first I didn't really think about it all that seriously. For one thing, I had spent the preceding twenty years of my life in medical schools. You know: get paid more and don't teach.

But by then I had been given ten positions to fill at Davis and had filled them all in a little over three years. I was in my prime, and I wanted to do more. I was itchy in my usual sort of way, and I longed for New England. Our weekends in Tahoe kept snow, trees, skiing, and log fires all very much alive in our personal lives. Still, I couldn't imagine going through another move, another disruption of the family, and another time sink. Somewhere along the way, one of my new nonacademic friends in Davis told me how executives move, or how they used to move. The corporation that wanted to move someone from place A to place B did the heavy lifting. They bought your house where you were, so you could have the cash to buy a house were you were going. It is called "executive hiring." I sat there listening to him and felt somewhat like a chump. Academics don't do that, I thought. He must be out of his mind.

When the next wave of nostalgia hit, I raised the idea of "executive hire" and almost instantly back came the reply, "We can do that." Armed with that, plus a couple of trips to check it all out again, we decided to move back to Dartmouth. It was all crazy, made even more so by the fact that the real estate agent found us a beautiful house in Sharon, Vermont, that inflamed the imagination, overlooking eighty acres of Vermont landscape with Mount Killington in full view. After Davis, everything looked green and lush. Right next to the house was an inviting pond to swim in, which in turn was next to a barn, which was fenced in with a twenty-acre post-and-rail fence. You get the idea. Oh yes, hiking trails all over the place. I took videos of the house, flew back to Davis, sat the family down to watch my new conquest, and we decided to do it. Of course, it is impossible to know what really kicks you over the decision line. It may have been that I missed not seeing and testing J.W. Or it may have been the tractor.

CHAPTER 8

STATELY LIVING
AND A CALL TO SERVICE

It would be possible to describe everything scientifically, but it would make no sense; it would be without meaning, as if you described a Beethoven symphony as a variation of wave pressure.

—ALBERT EINSTEIN

S OME PLACES DEMAND RESPECT FROM US. WE USUALLY think of institutions doing that, and when under their spell, we crank up our game as high as it can go. What took us by surprise was that our new home in Sharon, Vermont, had the same effect on us. In many ways, we had no business owning and living in such a stately place. It was situated on a plateau called Point of View, which the early settlers of our little town had so dubbed because of its 180-degree views of the Green Mountains. As we stood outside our back door, we could see Mount Ascutney straight ahead and, to the right, the Killington and Pico ski areas. The lawn swept down to a barn and pond (Figure 44). Soon after moving in, Antonio and Hannah Damasio visited us, stepped out to the back, stood there

FIGURE 44. Our magnificent home with a view in Sharon, Vermont. We shared countless memories here.

for a moment quite stunned, and then turned to us and said, "Why would anybody ever leave here?"

Even though we would live there for ten years, it was always called the Phillips House. Ellis Phillips, heir to the Long Island Electric Company, and Marion Grumman Phillips, heir to the Grumman aircraft company, had built it and nurtured its glory, with fences, gardens, and trails. They were extremely wealthy and in fact had several homes. With the financial distress of the early 1990s, however, they decided to sell it. By the time we were looking, the price had dropped to $575,000. That was still outside of my range. Nonetheless, I was captivated and for more than the usual reasons. The house had a John Deere tractor with all the attachments. I have been addicted to tractors ever since my high school days, when I thought I was going to go into farming as a livelihood. The tractor loomed large in my mind.

The real estate agent, Sue Green, was a good friend and the wife of a dear colleague, the psychiatrist Ron Green. Recalling Vermont winters, the wisdom of buying a home some twenty-five minutes from Hanover, most of which was a gravel road, had me vacillating. The other options were in town, more staid, highly functional, but none had tractors. Finally, I geared up my courage to make a low-ball offer. I called Sue and said, "Sue, four hundred and twenty-five thousand, with the tractor." There was a pause on the phone and Sue finally replied, "Mike, I can't in good faith pass that along. Also, you should keep the tractor out of it. Make a separate offer."

I thought for a moment. I teetered the other way. The family had seen the house only via videos that I had brought home on a previous trip. I teetered back: I was tired of looking. I suddenly found myself saying, "Okay, four hundred fifty thousand, and five thousand for the tractor." She said she would see what she could do, but was not encouraging. Five minutes later she called back. "Mike, four fifty is fine, but Mr. Phillips wants seven thousand for the tractor." This did not seem at all crazy to me. I knew there was something about a man and his tractor. Of course, none of this would have been possible without the executive hire deal I had made. My home in Davis had not sold, and I was cashless. Dartmouth advanced me the agreed-upon price for the Davis house, and I was ready to do business. The Phillips House had sat empty for a couple of years and needed some tuning up. We offed the wallpaper, ripped up the skanky wall-to-wall carpets, installed hardwood floors, and redid the kitchen. In addition, an unfinished space over the garage had to be finished off for my office. I contacted the original builder of the house, Hank Savelberg, who lived in nearby Woodstock. Hank not only remembered the home; he had designed it as well. He came over in a flash, basked in his beautiful work and our praise, gave a price, and the deal was done.

A few months later, the moving van pulled in from Davis, and we were set up for a life-changing experience. Over the years, I had developed a ritual to smooth each move. I went to our new home a week ahead of the family and unpacked the furniture, put the kitchen in order, and put some shine on the place. My real concern was the pond. My daughter Francesca, and Zack, my son, had grown up with a big blue kidney-shaped pool that radiated relief from the hot Davis sun. They lived in that frog-free, crystal clear pool, diving off the diving board and floating around in see-through plastic rafts. How were they going to respond to a greenish, slightly opaque pond loaded with frogs, scum, and more? Charlotte and I had an idea. Francesca could bring three of her Davis friends with her, one of whom was a frog freak! She loved the outdoors, loved the natural world. Maybe, just maybe . . .

After the long flight from San Francisco, they landed in Boston on a very warm July day. They jumped into a car and a couple of hours later were pulling into our new driveway. It swept up a gentle hill past green fields on the left and a row of maple trees along the right, which extended to the front door. Everyone was excited. I had my fingers crossed that it was all going to work. The doors sprang open and the kids piled out of the car. We were still exchanging hugs when Francesca's friend, Kirsten, grabbed her bag and yelled, "Let's get our bathing suits on and go to the pond." Within five minutes all were trunked up and running down the path. One after another they leapt into the pond, boom, boom, boom: *all* of them. Kirsten immediately caught a frog, beamed at her accomplishment, and all was well. The kids swam and played for two hours.

Charlotte and I were relieved and happy, but we also had one other slightly important event on the agenda: My oldest daughter, Marin, was getting married right there in our new home in less than a month. Her husband-to-be was, get this, Charlotte's brother! How does that work? Well, being married twice does the trick. My first wife and I had four beautiful girls, including Marin. Charlotte and I had been together about twenty years by then and during that time there had been lots of family gatherings. Marin and Chris had fallen in love. It was as simple as that.

"How many people are coming to the wedding?" Charlotte asked.

Oh, about two hundred or so, Marin said. For Charlotte, that was no problem. For the next four weeks, the extended Smylie clan (there are nine brothers and sisters) and the Gazzaniga clan (a mere six) tackled the grounds and the house, preparing for the wedding. Our efforts only needed to be augmented by a tent in case of Vermont weather. The culinary needs were handled by Charlotte's Texas brother, Ray, king of Texas barbecue. She had asked him if he would prepare his famous, mouthwatering, mesquite-flavored brisket, which he slowly barbecues in refurbished steel drums for twenty-four hours. He cooked eighteen of them in Uvalde and air-shipped them for the big day.

With the preparations for the wedding, this house began to define our family. It wasn't long before Francesca's sense of music, planning, acting, and writing came together in what proved to be a ten-year enterprise: musicals in the barn. The summer musicals evolved from the audience sitting in the horse paddock to full-scale productions in the barn. By the age of fifteen she was running the event as a summer camp for children of the Upper Valley. I put my carpentry skills to work and finished off a stage inside the barn, installing spotlights and a sound system along with a stage curtain. By the time she was in high school and her production of *Aida* came around, the Dartmouth stage designer took an interest in her "Sharon Players" and helped her with the stage scenery. The two-week summer school for kids became the place to be. She began to charge for the school and was befriended by a local businessman, who counseled her on insurance matters, unemployment issues, and finally in investing her profits in a Roth IRA. Throughout her life, Francesca has always set clear goals and achieved them with intelligence and verve. She is now a Ph.D. in molecular biology.

Meanwhile Zack, with the boyish energy of a ten-year-old, and I were discovering the woods. We ultimately installed an Amish ready-built one-room cabin at the far end of the property. In the summer we could walk to it, and in the winter we skied, snowshoed, or snowmobiled to it. Zack had mastered the land, set up paintball war zones, and soon became enthralled with the outdoors. Of course, this was in no small part due to Charlotte's honoring her pledge to him when leaving Davis that he could join the Boy Scouts. When we arrived, however, we discovered that Sharon did not have a troop, so Charlotte became the first female Scout leader in Vermont. That was cool with the other Vermont woodsmen. They taught her how to throw an axe and hit a tree at fifty feet, scale up mountains, and everything else. Every summer until Zack was fourteen, they went off to Scout camp. Charlotte had her own tent and became a bit of local legend. Zack and his two other buddies rose to the level of Eagle Scouts in record time.

Sharon was having its way with Charlotte and me as well. The little tweaks we had done to the interior's stunning rooms and the remake of the kitchen, with its capacity to serve a crowd, if necessary, all captured not only our imagination but also the imagination of the hundreds of scientists we had to dinner parties throughout the years.

There was always an element of family life that seemed perfectly natural in Vermont. Anyone who walked into our home felt the warmth and beauty of the place. Once, Bill Buckley came to visit. As he walked in the front door, he spotted the grand piano in front of the bay windows overlooking the Green Mountains. He dropped his suitcases, walked over, and started to play. Francesca immediately picked up on it and sat down next to him, and before long, they were playing duets. After dinners with visiting scientists, we would all adjourn to the living room for coffee and cognac, and Francesca and Zachary would come down from their rooms and play a tune or two. Zack had learned the trombone and Francesca played the piano or sax or steel drum. To this day, they are not the least bit shy about speaking or performing around adults (or anybody else, for that matter). The house in Sharon was magic. And, as we shall see, the inspiration of the place may also have served to prompt me to accept duties that went beyond the laboratory.

RELAUNCHING IN HANOVER:
THE INTERPRETER II

I can't stand the daily routine of academic life. Departmental meetings rank right up there as, in my book, a complete waste of time. The reason they are tolerated as much as they are is that a lot of people do like to fill up their time with such "necessary, thoughtful decisions." These "important" decisions might include whether to add a unit of credit to a statistics course, or whether to promote somebody, or whether to discipline a student, or how many pencils should be bought and, in fact, are necessary. I simply didn't care about any of it, even though somebody has to. My solution to this ambivalence about faculty meetings, and it came to me early in my career, was: Don't go. To my colleagues' credit, they rolled with it. Soon my behavior became one of my descriptors: "Oh, Mike doesn't do meetings." And that was that.

Of course, this kind of behavior only works if you do something else for the group (and they realize it). I was good at bringing together talent and masterminding large research grants. Those efforts take inordinate amounts of time and require knowing not only the research but the intricate layers of politics that exist on both the local and national level. Dartmouth College, not known for a hard-driving research program, suddenly had a nationally recognized program in cognitive neuroscience that was buzzing. At one point, we were responsible for more than half of the total research funds brought into the college. We were up and running.

By then, J.W. was driving and no longer accompanied by his mother or wife; he would drive down the interstate to our labs at Dartmouth on his own. We were following up on all sorts of issues, not the least of which was the nature and underlying mechanism of the "interpreter"—that special left brain device that generates a story about why we do the things we do. George Wolford, a longtime professor at Dartmouth, had taken an interest in this idea. He and

Mike Miller, who had come with us from Davis, wondered if a very simple, tried, and tested probability game, devised for understanding the nature of decision making, would be handled differently by the two half brains. It was painfully simple in design yet utterly provocative in its implications.

Imagine fixating on a dot displayed on a computer monitor; the chore is to guess if one of two different words is going to appear. Simple as that. Meanwhile, the experimenter is doing a little manipulation of the words. In fact, one of the words comes on the screen 70 percent of the time. What is the best strategy for guessing if the goal is to guess correctly as often as you can? To give some context, when a rat is given this kind of problem, it learns which choice leads to a reward more often, and then chooses that one all the time. In that way, it's guaranteed a 70 percent rate of success. This is known as probability maximizing. What do humans do? We think we are so smart! We think that there is a pattern that we can figure out: We try to deduce what the exact sequence of the stimuli is, so we can guess the correct answer on every trial. That is to say, we try to figure out the actual probability that a certain word will appear each time. So if we know that a word shows up 70 percent of the time, we guess that word 70 percent of the time. This is called probability matching. This yields us correct answers only 63 percent of the time. We humans are always trying to find the pattern, the cause and effect, the meaning of stuff. In doing so, we find our bizarre uniqueness. Here is how Lewis Thomas put it years ago.

> *Mistakes are the very basis of human thought, embedded there, feeding the structure like root nodules. If we were not provided with the knack of being wrong, we could never get anything useful done. . . . The hope is in the faculty of wrongness, the tendency toward error. The capacity to leap across mountains of information to land lightly on the wrong side represents the highest of human endowments. . . .*
>
> *The lower animals do not have this splendid freedom. They are*

limited, most of them, to absolute infallibility. Cats, for all their
good side, never make mistakes. I have never seen a maladroit,
clumsy or blundering cat. Dogs are sometimes fallible, occasionally
able to make charming minor mistakes, but they get this way trying
to mimic their masters.[1]

We found that only the left, smarty-pants hemisphere would try to guess the probability. The right hemisphere took the easy route and maximized, acting like a big rat. There was quite a bit of excitement about this simple experiment, meaning that people were quick to challenge its interpretation. That is how it should be and should go on in the very lab that gives rise to an idea. Mike Miller took this observation to another level. When he learned that goldfish and other simple creatures didn't just maximize, but at times also did probability matching, he began to wonder about our initial, simple interpretation. Within the code of the scientific method, a scientist's job is to disprove a hypothesis. Whenever I came out of my office into the bullpen, the graduate students and the postdocs were always talking, always thinking about assaults on what we thought was true.

As Miller was stewing about it, another postdoctoral fellow, the talented Columbia-trained psychologist Paul Corballis, started to wonder if the right hemisphere had its own kind of interpreter, one that specialized in visual information. Miller decided to see if the difference in strategy used by the left and right brain was actually based on something simple, such as the kind of stimuli that were being used. Instead of using a word as the stimulus, he switched to faces—which face was going to appear? Detection of words was a left hemisphere specialty, whereas detection of faces was known to be a specialty of the right hemisphere. Maybe the right hemisphere would change its strategy and try to guess the probability of whether the next stimulus was Face A or Face B. That is exactly what happened. Suddenly, the right hemisphere appeared to have those fancy left hemisphere skills, too, and was probability matching. Now it was the left hemisphere acting like a rat and maximizing.[2] What was going on?

The distinguished psychologist Randy Gallistel had written eloquently about the whole phenomenon of probability matching and how deeply biological it all seemed to be.[3] Imagine an organism foraging for food that has to pick between two trees. One of the trees has a 70 percent probability of having fruit while the other has only a 30 percent chance. It would be natural for the organism to choose the one with the higher payoff for food. As the organism nibbled away at the tree, however, the probabilities would be changing and at some point that other tree would become the hot property; thus one would expect that a highly evolved organism would also have built into it the constant monitoring of the other possibility. In this light, guessing probabilities becomes a basic structure, and then, once a probability is noted, maximizing would be the appropriate smart response.

As is so often the case in science, while an initial observation remains true, the initial interpretation can be completely wrong. Mike Miller had dogged it and wrestled it to the ground. It now seemed as though both hemispheres had these basic mechanisms, and what was lateralized was the interpreter, that unique device that tries to make sense of our thoughts, emotions, and behavior. Thanks to his efforts, and after putting in lots of work, we understood the mechanisms of hemisphere function more clearly.

BIG SCIENCE, SMALL COLLEGE

Dartmouth still needed an extra push in building up our new Center for Cognitive Neuroscience. We had been working in antiquated space with no real modern labs available. The college had been aware of this for some time—in fact, for some twenty years. Finally, some action was imminent. There would be a new building for psychology, and cognitive neuroscience would get the entire fourth floor.

I was delighted, but I also knew we needed to move into the era of brain imaging. In order to even think about having the capability of brain imaging in the new building, the right kind of basement addition would have to be constructed.

I decided to mobilize the faculty and to make a bold push to the dean: establish a brain imaging center with a new MRI machine right there in the new psychology building. No psychology department in the world had an MRI of its very own in its own building. Dartmouth should be the first! The dean at the time was a biologist and was very friendly to the idea of making the field of psychology more biological. Still, we were talking big-time money for a small liberal arts college. In truth, I thought it would remain a fantasy.

Dean Ed Berger called me with his decision. While he couldn't meet the whole nut, he could contribute $450,000 toward the additional basement costs. I had learned from days at Davis, one always takes any largesse made available for a project, even though it might not be the whole amount. In fact, it rarely is. I also knew, however, that it was not logical to spend that amount of money on a basement and not put anything in it.

There was another issue. No one on the faculty had the faintest idea how to run an fMRI machine. While some of us had been part of studies that included brain imaging measurements, we didn't know how to actually manage these extremely complex machines and their environment. That meant getting the dean to approve a full professorship and a national search. We knew we needed a leader in the field, but at the time, we didn't have a machine or the firm commitment of one. In 1999, while all of this was going on, I was running the two-week summer institute in cognitive neuroscience. One of the guest speakers was Scott T. Grafton, a brain imaging expert up from Emory. His work was fascinating, and his sense of a scientific problem was always right on. He was a neuroscientist in his own right as well as being a technical expert in imaging.

That afternoon, I dropped by the Hanover Inn—one of the most idyllic spots in America. Scott had changed into jogging clothes, which alarmed me: I hadn't seen mine since the aborted climb on Mount Rainier. The idea flickered through my mind that he might not be my kind of guy after all. We sat on a couple of cane rocking chairs on the front porch of the inn, overlooking the college campus that Dwight Eisenhower once noted looked exactly like what a college should look like. After a few thises and thats, I blurted out, How would you like a job here? He looked at me and simply said, "Why not?" I asked if his wife could come up for a visit in the near future. She was a general surgeon specializing in oncology, which meant we were talking two jobs. While I was not dissuaded, I was beginning to see hiring complexities. Although Dartmouth was small and purportedly under the aegis of one administration, the College of Arts and Sciences had little to do with the medical school and hospital and there would need to be a job open.

Two weeks later the Graftons pulled back into town to take a serious look at moving to Hanover. It took about two seconds to realize all would work out. Kim is one of those magical people. After she met with the surgeons, she was offered a job on the spot. Ironically, it was the college that had to go through some paperwork to get the offer out the door to Scott. Still, it was all done in record time, houses were bought and sold, and by New Year's Eve of 1999, the Graftons were in town to greet the spanking-new year and the spanking-new scanner.

The presence of the machine led to a change in the whole intellectual level and activity of the place. Having Grafton, a true authority at the helm, meant we were instantly taken seriously throughout the brain imaging community. Postdoctoral fellows from all over the world flocked to Dartmouth. Hiring new assistant professors became much easier. New kinds of funding were made available. The place was hopping. Again, it all worked because of Scott Grafton. Not only did he understand all the math, the physics, the computer science, and the data analysis issues, but at his core he was a psychological scientist. He wanted to know about how the brain plans action, perhaps the key question of all cognitive neuroscience.

Yet Scott had still another dimension. He was an M.D., a neurologist, working in a Ph.D. environment. Prior to switching to basic research full-time, he had practiced medicine for twenty years. He had walked the wards, declared people brain dead, seen suffering, treated all the patients that walked in the door, and done all the rest that goes with medicine. Dealing with the bumps and squeaks of a psychology department were only minor annoyances in comparison and simply didn't cause him any major anxiety or concern. His equanimity was almost unheard of and was widely appreciated.

What came with this posture was also a huge appetite for ideas and a willingness to help novices learn how to do complex brain imaging. As a result, when a social psychologist approached him about examining the multiple dimensions of the self, or the workings of the emotional brain, or the possible pathways of the brain involved in transferring a visual image from one area of the brain to another, or any one of dozens of other projects, Scott was there to make sure the science was done correctly.

CORRECTING SCIENTIFIC ERRORS

The last time I had used a scanner was in New York and it was to examine the extent to which our patient V.P. was truly split. That study was done with an earlier version of an MRI magnet, something called a 0.5 Tesla machine. It provided remarkable images, and, at the time, we swooned. At Dartmouth, our new machine was 1.5 Tesla in strength, which meant the signals being captured from the brain tissues were clearer and more detailed. Today, the everyday machines are 3 Tesla, and the experimental machines for humans are up to 7 Tesla. The stronger the magnet, the greater the signals, allowing for crisper images and more anatomical detail.

When V.P. came up to Dartmouth for one of her testing sessions, we thought it might be good to rescan her. We wanted to double-check her images to see if the fibers that we thought had been spared in her split-brain surgery had indeed been spared. For several years we believed that the surgeon had missed some fibers in the posterior region of the callosum, thereby possibly allowing for some kind of visual information to be communicated between the hemispheres. She also had some fibers spared in the farthest anterior regions of the callosum. No one knew what these regions might be communicating.

A few years earlier, Alan Kingston had come up with a dramatic finding that suggested V.P. had some unique capacities. When compound words were presented to J.W. and V.P., they responded differently. There was no ambiguity about remnant fibers in J.W. His postoperative MRI was clean as a whistle. When he was presented with a compound word, neatly separated, such that *sky* was flashed to the right hemisphere and *scraper* to the left hemisphere, J.W. drew a picture of scraper (serrated knife) with his right hand and a picture of the sky with clouds with his left. There was no integration, no picture of a tall building. V.P., on the other hand, integrated the information on every trial (Figure 45).

With this knowledge and because we were sure that those early imaging results were not only cool but accurate, we began to study what kind of visual information might be transmitted over those

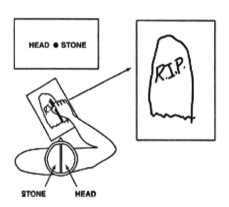

FIGURE 45. We tested Case V.P.'s ability to integrate visual information. Because her callosum fibers had been inadvertently spared during her surgery, she was able to do this in a unique way.

remaining posterior fibers. We assumed from all of our earlier work that the posterior fibers were doing the work. In poking around on this problem, Margaret Funnell came up with another puzzling result. V.P. seemed as split as anyone for 99 percent of the tests we ran on her. She could not cross-compare color, pattern, size, or anything else we could think of between her two hemispheres. Then, one day after trying various combinations of stimuli, Margaret presented a phrase to one hemisphere, "red square," and its corresponding colored geometric shape to the other hemisphere. Specifically, Margaret flashed the words to the right brain and, about a tenth of second later, a picture of a red square along with another shape to the left brain. Thus the task for V.P. was simple: After one half brain saw a word pair, such as "red square," all the other half brain had to do was to pick out a red square instead of the alternative, say, a blue circle. On this sort of trial she responded correctly. What was so mystifying was that for the comparison task, one of the half brains had to have the comparison stimulus printed out as a "word." If we flashed an actual red square, and not the words "red square," V.P. could not do the simple task! We were flabbergasted. Could it be possible that some of those remnant posterior fibers were selectively transmitting information about words?

My graduate student friend from Caltech days, Charles Hamilton, had been carrying out some intricate work on the callosal system in monkeys. Chuck had been showing that the posterior region of the callosum was segregated into areas that seemed to subserve different aspects of the visual experience.[4] It was intricate and fascinating work, and we quickly believed that with V.P.'s results we might have stumbled on to a homologue* in humans, pathways not directly related to simple sensory experience, but specific pathways dedicated to higher-order information. We wrote

---

* A homologous region is one that has the same evolutionary origin as another but may differ in function.

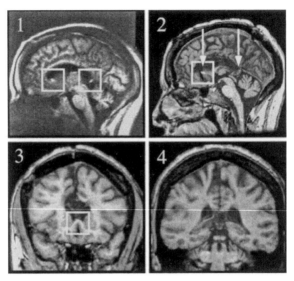

FIGURE 46. Images taken from MRI scans of the patient taken in 1984 and 2000. The white squares mark regions of bright signal observed at both ends of the corpus callosum in the 1984 scan (panel 1) and at the rostral end in the 2000 scan (panel 2). The arrows in panel 2 indicate the locations of the coronal slices shown in panels 3 and 4. Panel 3 shows a slice through the region of bright signal found in the anterior callosum, where the spared fibers can clearly be seen. Panel 4 shows a slice taken from the posterior end of the callosum, in the region where bright signal had been observed in 1984. The callosal fibers in this slice are clearly severed.

this up and gleefully sent it off for publication. It was quickly accepted and published.[5] It was after this that V.P. was scanned by the new machine.

You guessed it. The new scans told a different story about V.P. Those pesky remaining fibers in the splenium—the visual areas of the posterior callosum that we thought we had seen—were gone! It was the fibers in the anterior segments of the callosum that were crystal clear in the new images. The earlier image had produced an artifact, which we wrongly interpreted as remaining fibers in the splenium. We immediately put out a second, short article correcting the earlier claim (Figure 46).[6] In fact, with the correction, the results were just as intriguing. Now, the part of the brain that we know encodes complex information, the frontal lobes, was actually doing the communicating, not the posterior sensory regions. In other words it wasn't a carbon copy sort of thing where the basic sensory regions of one half brain were communicating with the basic sensory regions of the other half brain. Instead, some kind of more abstract representation was being communicated.

GETTING AHEAD OF A SOCIAL NORM

When I returned to Dartmouth from UC Davis, I had the good fortune to get to know a professor of mathematics, Dan Rockmore. Harvard- and Princeton-trained, and restless like me, Dan introduced me to the world of computer science. I had my Mac and that had been good enough for me, but Dan is one of those people who can't help themselves. He understands stuff, lots of stuff. He was a bachelor at the time and so had extra time to hang out and talk about ideas and possible new projects. He could easily understand my thinking, and I could trust he knew what he was talking about. Together we launched many new ideas.

One day we were hanging out at the Dirt Cowboy, the local coffee shop, when I told him we needed a database for all the brain imaging experiments being done all around the world. Brain imaging experiments are expensive, and the data generated from them could be used and reused if people shared their data. Scientists are always thinking of different ways to analyze a data set as time goes by. And making them available to anyone interested in the topic would be extremely productive. This was a well-recognized issue, and one of the leading researchers in the brain imaging field, Marc Raichle, was among many pushing for it. A National Institutes of Health official, Steve Koslow, was trying to identify funds for what would be called a neuroinformatics program. The highly influential editor of *Science* magazine, Floyd Bloom, was a very active supporter. There were many, many more and yet no one was doing anything about it.

Rockmore began to unpack the problem and explain how he thought it was a doable task. For a neuroscientist, the amount of data in any one experiment seemed huge and overwhelming. It caused a sort of paralysis in thinking and action. How could giga-bytes of data be managed? I mean, the computers would have to store maybe a terabyte! Holy cow. Over the next few weeks, Rock-more reinforced my salutary view of interdisciplinary collabora-tions. To him and his colleagues, large data sets were no problem at all. He soon brought in other mathematicians and computer scien-tists. Before I knew it, we were submitting a large grant to open up a national functional brain imaging data center (fMRIDC). Raichle agreed to serve as chairman of our external advisory board, Koslow said he would keep an eye on the application as it wound its way through the government review bureaucracy, and we started the spadework. We thought we were heroes, especially after we were actually funded by not only the National Science Foundation but also the Keck Foundation.

Of course, achieving that first goal involved a lot of footwork. First and foremost was the huge sociological problem that scientists do not like to share their data. Turf is king. At that point in scien-tific history, physicists, astronomers, geneticists, protein chemists, and more were commonly sharing their data, but neuroscientists had never been asked to do so. None of the disciplines that were cur-rently cooperating had liked the idea at first. They all fought it until the norm in the field switched, usually led by one of its intellectual leaders. Quickly enough, the journals in the field would only publish a paper on the condition that the underlying data from the paper was made publicly available. This whole process was not always swift, and we had to submit our grant within months. How were we going to convince the grant agencies that researchers would hand over their data?

Conveniently, I was editor in chief of the *Journal of Cognitive Neuroscience*, so I decided the journal should have a new manda-tory policy about data submission. In order to publish in *JOCN*, you would be required to submit your data to our new database system

Naturally, we wrote to all the major journals and asked for the same requirement, and all agreed at first.

As the project rolled on, it generated a lot of heat. As my colleague Jack Van Horn, who managed the project with great skill and care, recently noted:

> . . . [U]pon becoming aware of our efforts and goals, fMRI researchers angered by journal requirements to provide copies of the fMRI data from their published articles began a letter writing campaign seeking to muster opposition against the fMRIDC—an effort which was featured in the news and editorial sections of several influential journals (Aldhous, 2000; Bookheimer, 2000). Commentaries over fMRI data sharing were aired in the pages of Science (Marshall, 2000), Nature (Editorial, 2000b), Nature Neuroscience (Editorial, 2000a), as well as in the journal NeuroImage (Toga, 2002) expressing concern over the data sharing requirement, over what possession of the data implied, human subject concerns, and, if databasing was to be conducted at all, how it should be conducted "properly." Leadership groups in the field argued that fMRI was not mature enough to begin archiving its data (Governing Council of the Organization for Human Brain Mapping, 2001), conjecturing that until the BOLD response* was better understood, it was too early to consider databasing the images of published studies. People privately complained away, that those who collected the data owned it, that they would be remiss in simply giving it and that a small team at a modest Ivy League institution was not the best group to take on the task of archiving it. As a result of the apparent magnitude of these concerns, many of the journals,

---

* BOLD (blood oxygen level dependent) contrast is the measure used in functional magnetic resonance imaging that relies on intrinsic changes in hemoglobin oxygenation between the arterial and venous blood. Everyone uses it and has from the beginning.

who had initially been so supportive, decided not to require sub-
mission of data from the fMRI studies they published. Instead
they hoped to wait out the controversy and let the community
itself resolve the issue.[7]

In retrospect, none of this is surprising as the human comedy goes, even with scientists. Nonetheless, the Dartmouth project continued onward for several years, thanks to the diligent work of Van Horn and many others. Koslow, who was at the NIMH, somehow had arranged for the initial funding of the project through the National Science Foundation. He approved of its progress and renewed the funding for the project for five years in 2004. Unfortunately, he left the NIMH, and the new project leader cut the funding to two years, and that was that. It was too bad, since thousands of people used the database, both for research and as a teaching tool throughout the world. Today, multiple neuroimaging databases are emerging. The most noteworthy among them is the Human Connectome Project, an NIH-sponsored international study to map the functional and anatomical organization of the human brain. They are building upon the pioneering work accomplished at Dartmouth. All of this project was possible because a mathematician said to a neuroscientist in front of the Dirt Cowboy, "Oh we can do that."

GIVING YOUNG SCIENTISTS FREE REIN

Ironically, the college that gave rise to *Animal House* and to other male excesses has also been a place that has nurtured and developed young women. It all started to change for the better in 1974 when the college went co-ed, against the protestations of many. John Kemeny oversaw the transition. As with many such revolutions, Dartmouth alums now wonder why it took so long. There is something very cool about the women of Dartmouth. They get it, and "it" is the zany nature of Dartmouth life.

Conan O'Brien gave the commencement address in 2011. He is very much a Harvard man and is famously spunky, seemingly irreverent, and always funny. After poking some fun at the graduating class and standing at its tree-stump podium, he said:

> *Your insecurity is so great, Dartmouth, that you don't even think you deserve a real podium. I'm sorry. What the hell is this thing? It looks like you stole it from the set of Survivor: Nova Scotia. Seriously, it looks like something a bear would use at an AA meeting.*
>
> *No, Dartmouth, you must stand tall. Raise your heads high and feel proud.*
>
> *Because if Harvard, Yale, and Princeton are your self-involved, vain, name-dropping older brothers, you are the cool, sexually confident, lacrosse-playing younger sibling who knows how to throw a party and looks good in a down vest. Brown, of course, is your lesbian sister who never leaves her room. And Penn, Columbia, and Cornell—well, frankly, who gives a shit.*[8]

The men and women of Dartmouth roared their approval, as did the forty-first president of the United States, who was there to receive an honorary degree. Dartmouth women *are* cool in their vests. One of my seminar students back in 1998 was Sarah Tueting, a goalie for the championship Dartmouth women's hockey team. She went on to win two gold medals at the Olympics. Off the rink she was intensely interested in neuroscience but also calmly nonchalant. My class butted up against her practice time, so she was obliged to bring her hockey stick to class in order to quickly exit for the rink. She would slip into the seminar room, slide the stick on the table with her books, and then ask a penetrating question about the nature of consciousness. As Conan says, very cool.

Of course, the college is co-ed now at every level of education, including graduate students and postdoctoral trainees. My own laboratory was managed by a new Ph.D. from Dartmouth, Margaret Funnell, who had been the student of the distinguished memory psychologist Janet Metcalfe. As I was returning to Dartmouth from Davis, Margaret had written to ask if she could test J.W. on a memory test. I had known Margaret from my earlier time at the medical school, when she was interested in speech pathology. I immediately wrote her back and offered her a job. It was one of my better decisions. Before I could blink, she took over my program and ran it effortlessly and to everyone's delight. Margaret's husband, Jamie Funnell, was soon to become the headmaster at Cardigan Mountain School, a residential school for young boys. Indeed, they dined with the students every day and, as a consequence, there was nothing they didn't know about male behavior. For Margaret, the lab was a source of amusement and, with her razor-sharp mind, a place to be creative with her own research and insights.

It was Margaret who noticed something odd about the response of J.W.'s left hemisphere to a simple perceptual test. She had projected two objects, one above the other. The only difference between the two was their orientation. All each half brain had to do was judge if they were oriented similarly or not. Unbelievably, J.W.'s left half brain, the language-dominant half brain, could not do the task, while the silent right half brain performed perfectly.[9] This simple result launched a vigorous research program that, in the end, yielded an important new insight developed by Margaret and her close colleague Paul Corballis. As I hinted earlier, along with the left hemisphere's interpreter, they were discovering a right hemisphere interpreter for visual information. Think about that—there is a special, lateralized process in the right half brain that gives us the capacity to judge whether two visual objects are oriented in the same direction or not. The left speaking and analytical hemisphere, if disconnected from the right, cannot do that simple task. On a larger canvas, it points to the fact that just because a half brain can see, can categorize, can spell, can name, and can associate, it doesn't mean it can judge orientation. Orientation uses a different module, and in humans, it has taken up residence in the right half brain. To use the current vernacular: It is amaaazing.

Next, Abigail Baird roared, and I do mean roared, into our Dartmouth lives. A freshly minted Ph.D. from Harvard and a Vassar graduate before that, Abigail was like nothing I'd ever seen before. It becomes boring to say things like she was bright, energetic, and all the other descriptors that apply to everyone mentioned in this book. Abby was formidable and very funny. In untypical fashion for an aspiring academic scientist, she bought an old house about twenty miles outside of town and single-handedly remodeled it from top to bottom. She frequently came to the lab in her work clothes, splattered with paint and plaster but always ready to jump in and do the science. She launched a very clever research program using fMRI on something we adults see as very curious: the teenage brain.[10] She was one of the first people to argue that teenagers' brains were not fully wired up. Before I knew it, she was being recruited away from my program to be a member of the Department of Psychological and Brain Sciences as a social psychologist in a regular tenure-line position. A powerful personality, Abby began to long for Vassar and returned to her alma mater after a few years.

During this period, a new graduate student joined the lab from Wellesley, Molly Colvin. She had her pick of graduate schools and chose the woods of New Hampshire, in part because of the vibrancy of our brain imaging facilities and faculty. There were still no other psychology programs that had their own scanner, and none of them had Scott Grafton. She too was a pistol. So now I had Margaret, Abby, and Molly, not to mention several of the undergraduates all making our science better, and the social scene more organic and sophisticated. One undergraduate working in our lab at the time, Megan Steven, a boxer, became a Rhodes Scholar and went off to Oxford to work with the famous neuroscientist Colin Blakemore. She wound up studying synesthesia,* becoming one of the first to discover

---

* Synesthesia is a neurological condition in which sensory stimulation from one sense or stimulation from one cognitive pathway leads to automatic, involuntary experiences in another. For instance, hearing a particular word may be experienced as a particular taste.

how the brains of people with this condition differ from the brains of ordinary folks.

Advances in neuroimaging were occurring fast at the turn of the twenty-first century. Not only was fMRI able to detect where various cognitive processes were occurring in the brain, but new measuring techniques were adding the neural tract information, or put more simply, how information gets from one active place in the brain to another. Putting the "place" and "connection" techniques together was one of imaging's recent accomplishments. This got us wondering: If we asked subjects to do a simple perceptual task that required processes known to be lateralized to one half brain or the other, then could we see the two different sites of activation *and* the activity of the neurons that must somehow coordinate the two sites? If so, maybe we could also capture the neural processes underlying the individual variation that is commonly seen in any sort of simple behavioral task. In other words, some people do simple tasks quickly, and some take more time to do the same thing. Were different pathways active for the fast versus the slow responders? Abby, later joined by Molly and Megan, studied several dimensions of this research question and found that this latter is the case.[11/12] The idea is that the individual variation in response times seen in any group of people is correlated with discrete and different neural pathways. The fast responders take the first cutoff—via the neural fibers closest to the sensory areas of the brain—to the other hemisphere while the slow ones take a later path.

When there is a good buzz in a lab, everybody gets in on it. Soon enough, Matt Roser, a student of Michael Corballis's,[*] Paul's famous father, came to us from New Zealand. David Turk, a student of the equally famous Alan Baddeley,[†] joined the lab from University of

---

[*] Michael Corballis is a psychologist at the University of Auckland who, among other things, studies the origins and evolution of human language and proposed that it evolved through gesture.

[†] Alan Baddeley is a British psychologist, well known for his research on working memory.

Bristol, and Todd Handy came back to us with his freshly minted Ph.D. from Davis. One of the side benefits of hiring such incredible talent is that their mentors come and visit and also join in the fun. Life was very good indeed.

IMPORTANT INTERLUDES:
SERVING ON THE PRESIDENT'S COUNCIL
OF BIOETHICS

Like most Americans, I went to work on September 11, 2001, with no special issues to think about. I was actually leaving that night for a trip to Germany when word came in about the World Trade Center attack. The first version was that a small plane must have hit it, and the plane was demolished. Curious, I thought, but not earthshaking. My wife and I had spent many wonderful evenings in the restaurant, Windows on the World, including our wedding lunch. Indeed, in the spring of 2001 our beloved Cognitive Neuroscience Society meeting was held at the adjoining hotel, and we had a reception at Windows as well.

It took only a few minutes to get the news straight. Before I knew it, the grad students had figured out how to hook the sophisticated video teaching room up to CNN. Soon a couple of dozen of us were staring at the scene aghast, many weeping, many simply stunned, as the realization hit, that the building was the least important issue. Then the second plane hit, and soon after, the first tower collapsed. Now there was chaos. I grabbed Scott Grafton, and we went to my house and sat frozen on the sofa watching TV as that horrible day unfolded. As much as we tried to abstract the situation, it couldn't be done. Like most Americans, we were deeply affected.

It is hard to capture the strong feelings of patriotism that ignited Americans after the September attack. In the following days, every-body I knew, independent of their personal politics, wanted to do something. Americans—and most of the free world—were pissed off. So, when about a month later I got a call from Leon Kass,* I was open to possibilities. He introduced himself on the phone and explained that he had been appointed by President George W. Bush to head up a bioethics council to deal with upcoming biomedical technological advancements. Would I be interested in joining the council? I didn't know what to say other than an immediate "Yes." At the same time, I didn't know much about bioethics, and I wasn't sure they had the right man. Kass assured me the council was about bioethics, but not composed entirely of bioethicists. At the time, there was no discussion about who the other members might be, how he got my name, my political affiliation or beliefs, and the extent of my knowledge about immediate issues such as stem cell research. Bush had made a speech about stem cells that August; while I had heard it, I didn't think much about it other than it seemed balanced to me at the time. Quite frankly, like most busy people, unless the topic was right up my alley I simply nodded at it.

An incredible process unfolded after that call and before the actual meeting in Washington. The White House Personnel Office starts to vet, as does the FBI. There are endless forms to fill out, including assurances that there are no conflicts of interest with investments and other outside commitments. Friends and neighbors are called about your character; guarantees are given that you have no undocumented workers or others on a personal payroll for whom you are not paying Social Security taxes. It really is quite unbelievable.

Knowing that the first meeting was coming in January 2002, I began to bone up on stem cells. What was so interesting to me was that while everyone seemed to have an opinion about them, very few people knew the first thing about their underlying biology, even colleagues in biology. We all had a sort of introductory biological knowledge, usually vaguely stated. What was the big deal anyway

---

* A University of Chicago professor who has been engaged in ethical and philosophical issues raised by biomedical advances.

with this new technology? Then the penny dropped. It's all about the embryo question, the question of when human life begins. Or really, the question is: Is there a difference between when life begins and when life as a human begins? When does one confer all the rights of a born human on a multiplying group of cells? Boom, it was the scientific/political question of the century. The meeting was going to be the place to discuss this grand issue, and I had no idea it was going to draw me into an eight-year investment of time.

When the first meeting finally occurred in Washington, I met the other seventeen members of the council who took part in the stem cell vote. I knew many of them by reputation, but only one of them personally. Paul McHugh had been the chair of psychiatry at Johns Hopkins for years and is credited with introducing biological psychiatry to American clinical medicine. Though psychoanalysis was still prevalent, he thought that by better understanding the brain, most of mental disease would be better understood. He is a remarkable human being in every way, and I adore him. He is a Democrat, a Catholic, and unpredictable in so many delightful ways. With his Boston brogue and sparkling glint, he could slay a tiger or gently turn aside an aggressive remark. After all, he was an experienced psychiatrist and had seen everything.

The first day of the council meeting in January 2002, we all had our moment in the sun and gathered together in the Roosevelt Room at the White House. We were to get our marching orders from President Bush, and all of us were eager and attentive. The president came into what is a tiny room, took charge of the meeting, and encouraged us to leave no stone unturned in our discussions. He then mused, "I like debate, and let me tell you, you haven't seen debate until you have heard Rumsfeld and Powell go at it."

The president then asked each of us to say a couple of words about who we were and what we did. It started out very formally with a string of statements in the form of "I am Professor X from Harvard and I do Y." It finally came time for Paul to speak and I will never forget it. Paul said, "Mr. President, I am Paul McHugh, and before anything else, let me ask you how are *you* doing?" A few days before, it had been widely reported the president had been watching a Sunday football game and had slipped off the couch and banged his head, causing a gash above his eyebrow. Bush broke into a big grin and said, "Well, other than feeling pretty silly about falling off a couch and finding myself looking up at my dog, I feel good. I have never fallen off anything before—without drink." In one quick move, McHugh had broken the ice for all of us, and the president, after poking fun at himself, had set a firm but warm agenda.

In fact, the council was packed with talented people. It reflected a true cross section of the intellectual and political culture, and because of that underlying fact, it became a hot potato. Usually Washington councils dealing with issues such as bioethics or biomedical issues are one-sided, and all reflect the secular views of most modern academics. Utility and mechanism are discussed, not Aristotelian categories, concepts of justice, means/end, is/ought, and a host of other philosophical and now-political issues involved with human decision making. There were battles royal going on all the time, and yet Leon Kass kept it mainly civil.

ETHICS, EMBRYOS, AND POLITICS

As I look back at those eight years, it was all about the stem cells. It started to become clear to me after the January meeting where my thinking would lead me. The gravity of the topic found me talking about the issues involved—the beginning of life, the idea of abortion—the issue of, as the Yale surgeon Richard Seltzer put it, "life avulsed."[13] At work, at professional meetings, and at the dinner table, the embryo question always generates reactions. One evening I brought the topic up with Francesca and Zachary. At the time, Francesca was a high school biology student and already had her views, driven by her understanding of cell processes. She wanted to start

STATELY LIVING AND A CALL TO SERVICE | 315

a national science-club movement called the "totipotes." When my son was asked, "When do you think life begins?" he matter-of-factly rejoined, not looking up from his enormous plate of food, "Not until your first open-field tackle."

Beyond a surface familiarity with the idea of stem cells, I really didn't see all the important distinctions. I was at the same level as most people who thought about it (that is, when they thought about the issue at all): Stem cells can help cure disease, but when they come from embryos that will then be destroyed, it pisses people off. While I didn't know much, I did know whom to call, my buddy Ira Black. Ira was a molecular neurobiologist and practicing neurologist who would soon be running the New Jersey Stem Cell Institute. We had been colleagues at Cornell, and he was a co-conspirator on dozens of projects. He was truly engaging, worked around the clock, and was always full of good cheer.

One snowy winter evening in Sharon, my wife and I called Ira, who was still at work, while we sat cozy in our den with a roaring fire. Ira was deep into his work on adult stem cells, which are different from embryonic stem cells and hold out a different promise for biomedical use. During that phone call, Ira laid out the story. Throughout that year, he became my constant consultant on what was going on with stem cell research.

As Ira explained to us that night after dinner, it goes like this. Normally, an egg and a sperm come together in the Fallopian tube* to form a zygote, which usually travels down the tube and implants onto the wall of the uterus within fourteen days. After implantation, it is referred to as the embryo. The processes that commence to germinate the nervous system begin after day fourteen. The embryo develops and differentiates, and about eight weeks after fertilization, it is referred to as a fetus. That, everybody sort of knows.

Complicating this is what is not so well known: Twinning commonly occurs during those fourteen days, and chimeras may also

---

* The tube that connects the ovary to the uterus.

be formed. A chimera comes about when two zygotes, the result of two different eggs being fertilized by two different sperm (fraternal twins), fuse back into a single zygote. The organism that develops can possess different sets of chromosomes in different organs! Still, the question is, once the sperm impregnates the egg, at what point is society supposed to confer on it all the rights of an adult human? Those who would bestow these at the beginning, that is, at the moment of fertilization and the appearance of the freshly minted zygote, are proponents of what is generally referred to as the potentiality argument. If the two-cell zygote were left alone (in its hostess, of course), it could be a human.

Ira went on to explain what this all meant in terms of stem cells. After the egg and sperm come together, the zygote divides into two cells, then four cells, then eight cells, then sixteen. All of these cells are called totipotent, which means any one of those cells could build the whole organism—a baby. This is what my daughter was referring to. As I said, she was well ahead of the game. As the cells keep on dividing, a subsequent stage appears, and that is called the blastocyst and is 70–100 cells in size. The blastocyst is a ball of cells with an outer layer and an inner grouping of cells. The inner group of cells consists of the much sought-after stem cells. They are referred to as pluripotent because, while they can't grow into a whole organism like the totipotents can, they can become any organ in the body, which is why they are wanted by biomedical scientists. Hearts degenerate, brains degenerate, lungs, kidneys, cartilage, you name it. The idea is to grab those cells and strategically place them in patients who have a specific organ disease. The new stem cells would be there to help mend the part of the body they are injected into. That's all you need to know about the biology if public policy is being decided. But that, as I found out, was only the beginning.

As I noted, lots of people on the council were Catholics. It would have been easy to assume that they would automatically be against stem cell research, because embryonic stem cell research did mean destroying the embryo, and that was against church doctrine. One doesn't have to go very far back in church history, however, to see that its opinion on this matter hadn't been settled until the late 1800s. The issue that captured the church was the issue of ensoulment and when it occurred during development. A church council decided to call it at conception, instead of the time frame St. Thomas Aquinas had argued in the thirteenth century, which was at around three months of gestation.

But none of that mattered. It was 2002 and what did the Catholic members of the council think? What did the Jewish members think, the secularists, the other Christians, the Republicans, the Democrats, the liberals, the conservatives, the women, the men, the scientist, the bioethicists, the humanists, the lawyers, the physicians, the . . . ? All of these professions and belief systems were in play and all were intently following the information being presented to the council. And the press was watching us, while we listened to all the expert testimony being presented on the nature of stem cell research, the nature and reality of what happens in normal routine, sexual union. It was intense and made my day job back at Dartmouth look like kindergarten.

It really was sink or swim as I began the process with no training. Suddenly, the feeling of having a point of view began to emerge. Sure, I hadn't thought about the issues in a deep way before, but that didn't mean I couldn't start to think about them then. As I learned on the council, thinking about moral and ethical issues is at the very core of what it means to be human. In many ways it was an awakening for me. Simply mentioning concepts was no longer acceptable. While the lights and cameras were on, what did you actually think about the serious business of running a society? What was the moral fabric going to be? Was massive stem cell research going to rent the very fabric of human culture?

As we all sat there listening to various experts over a six-month period, several telling points emerged. At the second meeting in February, Irv Weissman, the distinguished stem cell expert from Stanford and the chair of the National Academy of Sciences' new report on stem cell technologies, came and gave a presentation. We exchanged pleasantries before the session, and it turned out we were both in the Dartmouth Class of 1961. We hadn't met back then, as he had left after three months to go back to his beloved Montana. He was a warm and extremely sure-footed advocate for stem cell research.

The National Academy of Sciences (NAS) report[14] was an attempt to demystify the stem cell issue and to draw lines between various proposed processes, such as adult stem cell research, embryonic stem cell research, reproductive cloning, and something called "somatic cell nuclear transfer" (SCNT).* Weissman ran into a buzz saw right away, led by one of the council's moral guardians of the Christian faith, Gil Meilaender. Gil was a professor of moral theology from a small midwestern college and a delightful, dry-humored provocateur. His concern captured a lot of the tensions that occur between reductionists like Weissman and humanists like himself. He basically asserted the NAS report had used different terminologies for the same thing, the thing that was at issue: the human embryo. Here is what Gil asked Irv:

> The Academies [sic] report discusses two procedures which it says are very different from each other. First, human reproductive cloning and, second, nuclear transplantation to produce stem cells. Suppose we are shown externalized in the laboratory two cloned blastocysts X and Y. We are not told which is X and which is Y but we are told that X is the result of procedure one and Y

---

* Somatic cell nuclear transfer is a technique wherein the nucleus of a donor somatic cell (any body cell except a reproductive one or an undifferentiated stem cell) is removed. Then the nucleus of a host's egg cell is removed and discarded. The donor nucleus is inserted into the egg cell and that egg cell will reprogram it. The egg, now sporting the nucleus of the donor somatic cell as its own, is stimulated with a shock and will begin to divide, eventually forming a blastocyst with DNA that is almost identical to the original host organism's.

is the result of procedure two, and we are asked to examine the blastocyst[s] and determine which is X and which is Y. On what basis could we make that determination?[15]

It was this exchange that got me thinking about the paradox, indeed the abundant misunderstandings, between the two. Gil was correct at the biological level. The biologic entity, the blastocyst produced either way, could go on to be a human organism if implanted into a uterus. Each of the blastocysts could have had its stem cells harvested for biomedical research. That is a simple fact.

At the same time, the processes were completely different through Irv's eyes, because the intention of the person carrying out the process was totally different. Paradoxically, Irv the reductionist believed the blastocyst was nothing but a bunch of molecules lacking in any serious sense what a functioning human mental person entailed. A luscious heirloom tomato can be mashed up and put into pizza sauce or delicately sliced to be in caprese. What happens to it is all in the mind and hands of the chef. Similarly, the intention of the scientist was not to make a whole organism or more bluntly, a baby. It was to make lifesaving cells that could help a patient suffering from disease. At the time of the writing of the report, nobody wanted cloning for baby making. Scientists generally thought it was dangerous and potentially harmful. Others felt only God gave life, and that that process was not to be tampered with.

Clearly, the majority of scientists on the council and elsewhere viewed the blastocyst as a bunch of cells, and the moral question was centered on what was to be done with them. Gil's view was that it wasn't a bunch of cells; it was already human. It started to become clear to me that the issue to be understood, in a fundamental way, was not specifically, when does human life begin? It was: What moral status should be conferred to a blastocyst? Indeed, what does it mean to be human?

While those stirrings were going on, other experts kept on coming before the council. One of the most memorable was a gynecological researcher from the University of Utah. He provided data that was a showstopper to many. Thirty to 80 percent of fertilized eggs resulting from natural sexual unions were spontaneously aborted! During one of the breaks, one of the Catholics said to me, "My word, are women supposed to have funerals for those people?"

Other cracks in the armor were provided by some of the council members themselves. Michael Sandel, the famous political philosopher from Harvard, began to dissect the logic of President Bush's position on stem cell research as not making any moral sense. On the one hand, Bush had ordered that no federal funding could be used on biomedical cloning because of the sanctity of human life, and that meant no embryos could be destroyed. But on the other hand, as Sandel pointed out, he didn't object to biomedical cloning going forward with private funding. So, it's okay to kill when privately funded?

During the council meetings, I started in with various metaphors that began to form in my own reasoning. From my point of view, the parts are not the whole, especially when the brain is not yet a part. The analogy I came up with: "When a Home Depot burns down, the headline in the paper is not '30 Houses Burn Down.' It is 'Home Depot Burned Down.'" The parts in the store are just that, not whole houses.

I also tried the analogy of the widely accepted "brain death" argument used when considering human organ transplantation. Clinical criteria had been established for brain death that proved solid and reliable. When there was irreversible brain damage yielding a flat EEG, the organs, including the heart, could be harvested and transplanted to keep someone else alive. None other than Pope Pius XII had supported this position. I reasoned, if brain death was accepted as a concept that allowed for the organs to be used for health goods, why couldn't the cells of a brainless entity like a blastocyst?

I slowly realized I was becoming an advocate. I marshaled all of my arguments and wrote an op-ed for the *New York Times*.[16] By late spring the lines had been drawn. At the June meeting, each member of the council had a public moment to say what he or she thought about the issues that had been swirling around. It came down to a vote on what each of us felt about reproductive cloning and biomedical cloning. The pressure had been building so much that *Times* columnist William Safire in mid-May wrote an article about the impending split in the council's sentiments about cloning.[17] The reporters were all over us for months, and various views, prejudices, perspectives, and more were being offered to anyone who would listen.

Kass offered us a choice at the June meeting. He had prepared some recommendations that reflected all the options discussed over the previous five months, and he wanted us each to indicate where we stood. The two main options:

> *Possibility three, ban on cloning to produce children . . . but under regulation of the use of cloned human embryos for biomedical research. Option three called* **ban plus regulation***.*
>
> *Option six, a ban on cloning to produce children with a moratorium . . . being understood to be a temporary ban . . . with a fixed time period on cloning for biomedical research. Option six,* **ban plus moratorium***.*[18]

Each of seventeen members spoke up on their preference and provided reasons as to why. Everyone's view was clearly stated, even the three members who were struggling with their decision. It came down to this: All members were for the ban on reproductive cloning. While one could draw upon either scientific or religious reasons, basically everybody thought it was spooky and weird.

As I say, the June vote was clear. Seven voted to ban biomedical cloning as well, which is to say, this group wanted a moratorium. As they freely acknowledged, they could live with the term *moratorium* because they wanted the extra time to convince the world that biomedical cloning was simply wrong.

Seven others, including me, voted for *regulation,* which meant to go ahead but to put regulations in place. Thus, this group had no moral problem with the idea of biomedical cloning. And finally, three other members said after some back-and-forth that they were for the possibility of biomedical cloning as well. All in all, this meant ten were for biomedical cloning and seven were against it. All of this is right there in the public transcript of this meeting. I was ecstatic thinking that our six months of work had revealed a reasonable position that had involved a true cross section of society.

Leon had not been happy with the whole idea of voting a stance. He argued the council was a place to air ideas and that that should be its function. He was a true product of the University of Chicago. But Washington with its bottom-line philosophy doesn't work that way. Subsequent to the June meeting, we were sent a form requiring a vote and our signature and were requested to quickly fax it back to the White House. One month later, at the July meeting, our report and the tally was published (Figure 47). Magically, in the intervening month, the sentiments of the council were represented as follows: Ten were for a moratorium and seven were for going ahead with regulation. Same vote, different spin.

During the weeks after the June meeting there must have been lots of politicking between the swing voters and those seeking a total ban. The seven of us who were clearly for going ahead were not involved. We had decided and were set. No point in wasting time on us. What was achieved from June to July was to lump the seven who wanted a total ban into the moratorium group and to add to that group the three who wanted to go forward with regulation and to convince them to be happy with the term "moratorium" instead of "regulation." In this way, it somehow could be perceived that the majority of the council was for putting on the brakes. Here is how it was reported in the *New York Times:*

FIGURE 47. On July 11, 2012, Leon Kass released the report on human cloning in a press conference in Washington, D.C. The press conference was exhausting, since Kass not only had to address those on the committee who disagreed with his position, but also had to acknowledge the political realities of both the White House and Congress and, of course, the press.

BUSH'S BIOETHICS ADVISORY PANEL RECOMMENDS A MORATORIUM, NOT A BAN, ON CLONING RESEARCH

Cloning for biomedical research should not be banned outright, but rather prohibited during a four-year moratorium that would allow time for more public debate, according to a long-awaited report by President Bush's bioethics advisors. By endorsing a moratorium on research cloning, the President's Council on Bioethics put itself slightly at odds with Mr. Bush, who supports a ban on all human cloning experiments. In a dissent, 7 of the panel's 18 members went even further, recommending that research cloning proceed under government regulation. Even the majority was split, according to an executive summary of the report, a copy of which was obtained by The New York Times. "Some of us hold that cloning for biomedical research can never be ethically pursued, and endorse a moratorium to enable us to continue to make our case in a more democratic way," the majority wrote. "Others of us support the moratorium because it would provide the time and incentive required to provide a system of national regulation." As expected, the council—which spent seven months examining the social and ethical implications of cloning experiments—called for a ban on using cloning to produce babies that are genetic copies of adults. Despite the divisions on the panel, a senior administration official called the report "consistent with the president's core view, which is that all human cloning is wrong and should not be authorized." The official added that the majority opinion "clearly rejects the position of those who say we should ban reproductive cloning this year while authorizing research cloning."[19]

Of course, it wasn't consistent with the core view at all, and Kass heard about it. The council members were clear on the dodge. Gil Meilaender, writing a few months later in the *New Atlantic,* accurately noted:

On the actual policy question itself, the deep divisions were apparent. Ten members of the Council supported a moratorium on cloning-for-biomedical-research, and seven favored moving ahead with such research, though only after regulatory controls were in place. (One of the original eighteen members had resigned and not yet been replaced before the report was released.)

It was, however, possible for everyone to claim a victory, if such claims matter. Because three of the ten-person majority favored a moratorium but not a permanent ban on cloning-for-biomedical-research, its advocates could—and did—emphasize that a majority of the Council opposed a ban.[20]

It wasn't long after the July meeting that the media pretty much lost interest in the council. The report was generally dismissed and relegated to that large hole in Washington where most reports are placed. Still, the media attention surrounding the preparation and publishing of the report brought to light how cool it would be if the biologists could figure out a way around the moral problem. That actually happened only four years later, thanks to the Japanese molecular biologist Shinya Yamanaka. Amazingly, he figured out how to take any cell in the body and turn it into a pluripotent stem cell by running its development back in time.[21] No more blastocysts discarded, no moral dilemmas, only procedures to take any cell and turn it into a cell that can regenerate the type of organ cell a diseased patient might need. Science marches on steadily, and Yamanaka was deservedly awarded the Nobel Prize six years later.

The council itself moved on and tackled all sorts of issues for almost eight years. I became increasingly frustrated with the council's portrayal of issues as slippery slopes, which I feel were accurately summed up by coolheaded Steven Pinker in 2008:

The sickness in theocon bioethics goes beyond imposing a Catholic agenda on a secular democracy and using "dignity" to condemn anything that gives someone the creeps. Ever since the cloning of Dolly the sheep a decade ago, the panic sown by conservative bioethicists, amplified by a sensationalist press, has turned the public discussion of bioethics into a miasma of scientific illiteracy. Brave New World, a work of fiction, is treated as inerrant prophesy. Cloning is confused with resurrecting the dead or mass-producing babies. Longevity becomes "immortal-

FIGURE 48. Jim Kim, the new president of Dartmouth, presents me with an honorary doctorate in 2011 on the occasion of my fiftieth class reunion. President George H. W. Bush was also awarded a degree, and Conan O'Brien gave the commencement speech.

ity," improvement becomes "perfection," the screening for disease genes becomes "designer babies" or even "reshaping the species." The reality is that biomedical research is a Sisyphean struggle to eke small increments in health from a staggeringly complex, entropy-beset human body. It is not, and probably never will be, a runaway train.[22]

MOVING ON AGAIN

The public service instinct bit me once again and I served as dean of the faculty at Dartmouth. A search committee unanimously nominated me as their candidate. They gave only one name to the president for consideration. In hindsight, I realized that may not have been the best footing on which to start a new job. Also, my years of not attending faculty meetings left me without a certain skill set. To be a successful dean of faculty you must either have a fine-honed strategy for navigating faculty politics or full support from the top. I had neither. I lasted only two years. Happily for me, the University of California, Santa Barbara, asked me to return, and it looked like another golden opportunity.

Six years after I left Dartmouth, on the occasion of my class's fiftieth reunion, Dartmouth awarded me an honorary degree (Figure 48). My family attended the event, and my class, the Class of 1961, asked me to give a lecture after the ceremony on my science. What actually occurred has been indelibly written into my mind. I let my stunning daughter, Francesca, Dartmouth Class of 2007 and then a graduate student at the University of California, San Francisco, give most of the talk. As John Kennedy said, "the torch has been passed." One of the deans came up to me afterward and said, "Does Francesca want a job?" Needless to say, I do love Dartmouth, as does our family, and on that occasion, my brother and I, along with his two sons and two of my daughters, established an award for the top graduating scientist.

PART 4

BRAIN LAYERS

CHAPTER 9

LAYERS AND DYNAMICS: SEEKING NEW PERSPECTIVES

The beautiful thing about a new idea is that you don't know about it yet.
—CHARLES TOWNES

I WAS READY FOR A NEW BEGINNING IN 2005, AND WHAT BETter place to begin it than in Santa Barbara, where I had started my professional career? Forty years earlier I had bought property in the seaside town of Carpinteria, just twenty minutes south of the UCSB campus, and I still owned it. Now, with the good luck of a new position, I was all set: My new book was coming along, the climate was intoxicating, and moving into the home that I had designed and built with my own two hands forty years earlier was fulfilling. Adding to that, many of my family lived close by. A new benefactor, the Sage Publishing Company, had bestowed a major gift on the university to start a center for the study of mind, and life seemed very good indeed.

I have a large family, so after considering the price tag of UC's life insurance for a sixty-six-year-old man, I decided to buy my own insurance on the private market. It was a lot cheaper, and the required physical was easily done in my own home by a traveling health-care professional. A quick EKG, a vial of blood, and it was all taken care of, or so I thought. A couple of weeks later, the insurance agent called to say that our request for a term life policy had been denied. I protested as he explained their policy was not to say why. A few days later he called back and sotto voce said the words no man wants to hear, "Your PSA was sixteen."

My prior medical care was under physicians who worked at the Dartmouth-Hitchcock Medical Center in Hanover. One of the leading epidemiologists in the world was the guiding light, and his studies had suggested there was little or no value to PSA testing, a biomarker test that detects blood levels of prostate-specific anti-gen (PSA). The physicians who took care of me said the decision to take the test was up to me. Knowing why the epidemiologist had his view, I sided with his reasoning, and, for the important ten years prior to my PSA of 16, I had never been tested. It all has to do with base rates of occurrence and outcomes. With large samples of data, epidemiologists had determined that, on average, patients with elevated PSA scores did not do significantly better with vari-ous medical interventions. In other words, there is no reason to get the test because if you have it, treating it doesn't improve the outcome. On average. The problem, of course, is that the average is made up of individuals and some are on the high end and are possibly helped. I can tell you, when you have been anointed with prostate cancer, all you can think about is: When can they start the interventions? The *don't just stand there, do something!* urge trumps detached statistical reasoning.

My nephew and my namesake, Michael Scott Gazzaniga, is a urologist and has clinical and patient skills that are over the top. He talked me through the troubling news, even saying he would do the biopsy for me down in his Orange County office. I had just heard some horror stories about a friend who had gone into sepsis after his biopsy and had almost died. Others routinely complain about the uncommon pain of it all. The specter of surgery, of radiation, of hormone therapy, of suffering the horrible painful death of prostate cancer, all rushed to mind. Mike calmed me down and said he had done more than three thousand biopsies, had never had a complication, and further, there would be no pain the way he did it. That was the good news. The bad news was that he determined I had the crud, real bad.

Mike quickly arranged for me to see the fabled urology surgeon at the University of Southern California, Donald Skinner. Skinner was my age, and his reputation for magical hands in prostate surgery was widely known. I remember him sweeping into the exam room holding my various charts and scans, and he simply said with a big smile, "Boy, what did you do to deserve this?" He walked my wife and me through it all and then arranged for surgery the following month, after a short course of hormone therapy to reduce the size of the prostate before surgery. On the morning of the surgery, Mike came up to sit in on the procedure, as he wanted to see the magic hands of Skinner and, of course, to offer his support. Away I was wheeled by the resident out of the prep room. I said something to Charlotte, and before I knew it, I was waking up in the intensive care unit with all the nurses smiling and saying it went well. It was a moment both weird and wonderful. My abdomen had just been filleted and my prostate cut out, removed through a foot-long incision. Within hours and with their help, the staff had me up and walking around. Morphine helps, but so does the driving good spirit of family and the staff of a well-run hospital.

The next day Dr. Skinner checked in on his morning rounds to see how I was doing. He had his flock of interns and residents following him, as is always the case in teaching hospitals. I felt compelled to ask the blunt question: "How long do I have to live?" Skinner paused and looked at me and said very calmly, "I have put you back onto the normal death curve." Charlotte and I rejoiced. We had thought, given the nature of the preoperative biopsies, I had about two years.

Such life events do drive us to mental places that I have systematically tried not to visit: considering death in general and early death in particular. Of course, we know about death; we have had good friends die, we have had parents die. But these experiences bear little relationship to the moment you're on deck. It's you, nothing existential or overly melodramatic, and the jig is tangibly up.

My views are nothing out of the ordinary. The lights are on and then they are off. When they are off, you won't know it because you will be dead. There will be no missing of family and friends because you will be dead, so there is no need to fret about that while you are alive. You will be missed by others, but that is true of them as well, should they suddenly die. You will miss out on things you wanted to do, which is true, but what does that mean since you will be dead and won't know it? And, on and on. In the end, by working through all of this, death seemed less scary, less paralyzing. Just like the leaves that fall off a tree, life ends. In fact, these ruminations proved to be good preparation for an honor that fell to me out of the blue about a year after my surgery.

GETTING READY FOR THE GIFFORDS

In 2007, I was invited to deliver the Gifford Lectures at the University of Edinburgh, for the year 2009. I was both shocked and overjoyed. Jacques Barzun had described the Gifford Lectures as virtuoso performances and "the highest honor in a philosopher's career." The lecture series, more than one hundred years old, was charged by Lord Gifford to discuss natural theology as a science, that is, "without reference to or reliance upon any supposed special exceptional or so-called miraculous revelation." That part I could do! It also meant that I had two years to prepare, and I now had good reason to believe I would live long enough to complete the assignment.

The invitation to give the Giffords motivated me to pull together my own thoughts on the weighty issues of human meaning and the future trajectories of neuroscience. Such assignments lift one out of the day-to-day, out of the next task to get done, out of the next proxi-mate issue to solve. I could feel my brain creak and crack under the strain of trying to lift my experiences up to a new level. As Nobel Prize–winning psychologist Daniel Kahneman likes to point out, our brains are lazy and don't like to work too hard.[1] And as George Miller had famously shown and most entertainingly wrote about,[2] we have a rather limited working memory and can only keep a few items in it active at a time. I needed another level of abstraction, that is, single-word substitutes, nicknames, jargon for involved concepts, to make some room for stuff in my working memory in order to understand how all the bits in which I was immersed fit together. I needed the bigger picture. I can do it, I thought. I'm a big-picture guy. After all, abstraction is what we revert to when details overwhelm us.

Neuroscience was all about the relationships between structure and function. Two main schools of thought had formed about how they interact. Some held that the brain's structure is set early and that the shifting actions of any organism come about by the very brain actually changing. Others held that multiple fixed structures in the brain are differentially called to duty, which only makes it look like a changing, plastic system, which in reality it is not. "Find out which is which!" was the call of the day, and the secrets of how the brain does its tricks would be unraveled. More accurately, a fixed versus plastic brain was one of the many dichotomies that had been set up. Another had to do with the deep reductionist beliefs of most scientists. Reductionism is a philosophical stance. It holds that a complex system is nothing but the sum of its parts. By looking at its parts, the whole can be predicted, and by looking at the whole, the parts can be described. In neuroscience, this translates to: A produces B, which produces C, a nice linear view of the world and a good one to start with if trying to figure out the brain. "Study sim-ple systems!," and the simpler the better: sea slugs, worms, and, if studying primates, then study only the behavior of single cells. Rats and mice were allowed for some issues. And again, all of this was proffered with a heavy reductionist bent that held that everything about the brain and what it produces can eventually be understood by simply looking at what the electrons in the neurons are up to. Starting from cognition to behavior to systems to cells to molecules, like peeling an orange, science could get to the bottom of it, to the seed, so to speak. And, like the seed that eventually produces a tree that produces an orange, we could build up again to the brain. So from electrons, then molecules, we could build in a straight line up to cognition. It would all fit into place. I was raised on this view, and, in many ways, I believed it, and to some extent I still do. At the same time, there is this tug on the mind, and it screams, "That can't be it!"

What is so palpably true is that the toiling scientist is always aware of the limits of his or her own work and of the possibility of barking up the wrong tree. When an idea falls from grace, it usually isn't because the original proponents had never thought of alternative views. They are usually painfully aware of other possible views about the underlying truth. They choose a side in the hunt and stay with it as long as possible and sometimes even longer than that. This is what Kahneman would call the "sunk cost fallacy," where you have invested so much already that you feel obligated to follow through. It is not right or wrong. It is what we humans do. In the current era of human brain imaging experiments, thousands of scientists are committed to finding the places and/or networks that seem to be more active during certain kinds of cognitive states. All realize, however, that this neophrenology may not be capturing the essence of how the brain does its magic to make us who we are and how we feel. I have often thought a clever way to make this all more explicit would be to have scientists submit with their papers for publication a review of their own work. I am sure their reviews would be the most scathing. Not many scientists are chumps.

As life goes on and various experiments sink in and either take hold or are thoroughly rejected, the overarching view of how the mind/brain issues are to be framed changes. In many ways, it is not unlike walking down a path, a well-trodden path, and suddenly seeing something new. It has been there all the time, but for reasons of belief, or ignorance, or fatigue, or attention to another thing, the something is not seen. When a graduate student at the University of Chicago's economics department pointed out to his professor, who deeply believed in efficient monetary markets, "Sir, there is a hundred-dollar bill there on the path," the professor replied, "That is impossible." Our theories blind us all. Still, it was my turn to give the big picture a shot.

BRAIN PRINCIPLES AT A GLANCE

Almost seventy years of neurobiological research has taught us that the brain is not a bowl of random spaghetti with its wires randomly spreading and slithering around with each toss of the chef. It is a highly structured biologic machine managing a complex code of action that had been called an "enchanted loom" years before by Sir Charles Sherrington.[3] An organism "remembers" its evolutionary successes, in both its structure and function, from the toenail or claw to the liver, through its DNA. The brain is hardly an exception, just another one of those parts for which successes have been coded in DNA. Who wants to start from scratch learning everything? That is not a good survival strategy. Better to pass some basics along to get things up and running as fast as possible. The brain comes with lots of programs that ready us for life's challenges.

Sperry laid this all out with his work on neurospecificity: In research that he had launched and developed before his Caltech days, he showed how the neurons in the brain hooked up.[4/5] When that knowledge is carefully considered, it prepares one for all kinds of future insights about the brain. Babies come from baby factories with capacities. With every passing year, we learn not only how much one-year-old babies know, but also how much one-month-old babies know: a lot. The field of developmental psychology keeps driving back the age at which babies reveal their cards. A group of clever Hungarian psychologists have now carefully observed babies' eye movements in psychologically structured moments and have discovered that six-week-old babies already have a theory of mind about others and see the social implications of other's gestures.[6] It also looks as if babies come born with a passion for teaching others new information. Unlike the zebra fish and the dog, which have their own extensive sensitivities to social order and more, human babies seem to come equipped to teach others.[7]

This formulation, that the brain is prewired in many ways, has withstood the test of time, even though some have a reflexive belief and want the brain to be infinitely malleable. It is an admirable desire and is rooted in the American belief that anything can be fixed by external reinforcements. Many professionals in the field cling to the idea of, for the lack of a better word, endless mutability. Unfortunately, most neurologic disease has a deadening finality to it. The slack such disease produces is not automatically taken up by another part of the brain.

The other idea that is powerfully captured in the last seventy years of neuroscience research is that processes of underlying behavior, cognition, and even consciousness itself are highly modular and work in parallel. Like most complex machines, this camp believes that parallel processes are ongoing throughout the brain's operations and are intricately producing a unitary function. Intricate, integrative processing from super-modularity seems almost nonsensical at first glance. And it gets worse. While disrupting any one part in an actual machine disrupts its operation, taking out hunks of the brain can frequently have little effect on that machine's behavior. Some hunks are vitally important, and some frosting on the cake. What's going on?[8]

First, to grasp the push toward modularity, think about one of the more remarkable aspects of a split-brain patient's reality. Prior to disconnecting the two hemispheres from each other, the left speech hemisphere merrily describes everything in full view. Just like you, when a patient is looking at someone's face, they can see not only the right half of the face but also the left half, miraculously stitched up at the midline by the brain into a unitary whole.

Now here comes the unfathomable part. Recall that after their surgery, their speaking left hemisphere now only sees the right half of the world. When you ask them, "What has changed?" difficult as it is to believe, they say not much. Everything seems more or less normal from the perspective of the left speaking hemisphere's point of view. How could that be? Can you imagine waking up after brain surgery, only seeing the right side of space, and saying not much has changed? Wouldn't you say something like, "Ahhh, Doc, I used to see all of your face when I looked at your nose, but now I only see half of it. What's up with that?" Indeed, why doesn't the left brain miss all the other stuff that the right brain was doing when they were connected? While you think about that, here is a hint: You don't think about or miss thinking about the gazillion unconscious processes that are ongoing in your brain that produce the conscious awareness you continually and uninterruptedly experience emanating from your left hemisphere. In fact, you don't even know that they exist unless you follow what's going on in brain research. Following hemisphere disconnection, what has happened is that everything that the right hemisphere did, and still does, has now joined the ranks of inaccessible processes. They have been added to that category of things you don't miss and don't think about.

It is hard not to come to the view that the very appreciation of the right brain's sensory phenomenal sphere (that is, everything out there on the left side of space) is housed in, and local to, the right hemisphere. It is bound to the local, physical processors that we know are active in apprehending that part of the visual world. Local processing reigns, very local processing, and it underlies all of brain organization. And much of that local processing goes on outside the realm of conscious awareness. It's modular, pervasive, and fast.

THINKING ABOUT MODULARITY

Modularity is a product of our big brains. A general principle of brain organization is that the larger an area is, the more neurons it has. The more neurons it has, the more neurons it is connected to. There is a limit, however, to all this connectivity. If each neuron were connected to every other one, our brains would have to be twenty kilometers (twelve miles) in diameter.[9] Talk about a big head. The distances that axons would have to travel across the brain would slow the processing speed down to where our body's movements would be all out of whack and our thinking would be tediously slow and dull-witted. That fat brain would also require so much energy that we would have to eat constantly and then some. So, as the ape brain evolved and got larger and the number of neurons increased, not every neuron connected to every other neuron. This resulted in an actual fall in the percentage of connectedness.

Since the internal structure and connectivity patterns change as the proportional connectivity decreases, a high level of clustering occurs, which gives the overall system greater tolerance for the failure of individual components or connections. The local networks in the brain are made up of neurons that are more highly connected to one another than to elements in other networks. This division of circuits into numerous networks both reduces the interdependence of networks and increases their robustness. What's more, it facilitates behavioral adaptation,[10] because each network can both function and change its function without affecting the rest of the system. These local specialized networks, which can perform unique functions and can adapt or evolve to external demands, are known as modules.

Modules! Modules! Everywhere Mother Nature screams it at us. If something is useful and already there, and modularized, Mother Nature uses it and keeps on going. As pointed out by Andy Clark, the distinguished philosopher from the University of Edinburgh, Hod Lipson and colleagues at Cornell University showed that "the control of apt finger motion is not enabled solely by the nervous system but involves complex and essential contributions from the network of linked tendons."[11] In other words, since we evolved to become more dexterous, why not use the information that is already in the torque relationships established in the tendons of the hand? That way the brain only needs to supply a much simpler set of instructions in order to carry out the complex task of individuated finger movements. It's a command more like "Grab the cup" than "Okay, thumb-press down with the force of x for y amount of time, and you, middle finger, spread to the side. . . ." This would allow for a "much wider range of directions and magnitudes of fingertip forces" than would otherwise be possible for the brain to orchestrate alone. In this case, and this is the part that Clark loves to remind us of, "part of the controller is embedded in the anatomy, contrary to current thinking that attributes the control of human anatomy exclusively to the nervous system." It's not just your brain calling the shots for that Chopin étude; it's your fingers that are partly in charge.

Mother Nature doesn't reinvent the wheel after every few revolutions. And likewise, not only does the brain modularize and push tasks and minute instructions out of its central processors, but the whole cognitive system, including the working brain, the body, and the environment, will call upon information embedded in the other to carry out a goal or action.

And yet, there is something wildly unsatisfactory about such an opinion if one holds tightly to the simple linear view of brain function: Module A goes to module B which goes to module C. If this thinking were correct, it would seem to suggest that many, many cables are crisscrossing the brain in an effort to keep all of the modules up to date, which as mentioned above, would require that twenty-kilometer brain. The simple reductionist view would also require a final box where all the toiling of the parts would finally be deposited and coordinated at a single point, and voilà, conscious experience itself is produced. That model was already under fire fifty years ago from the basic split-brain studies. Why, we wondered, is it that cutting the corpus callosum, the largest communication line in the brain, instantly results in two fairly similar conscious entities, side by side enjoying a common body? Suddenly there are two final loci generating conscious experience? How could a simple linear model suddenly produce, with one slice of the surgeon's knife, two conscious systems, side by side? In short, the simple linear model where A produces B, which produces C was not supported by the split-brain findings. New, or at least different, concepts were needed to grasp the very phenomenon being studied.

A ubiquitous concept in science is the notion of emergence: that more complex systems arise out of relatively simple interactions. Biology arises out of chemistry, which in turn arises out of particle physics. Similarly the mind arises out of neuronal interactions and above it economic principles arise out of psychology. It is a very slippery concept. And it seems tangibly present, especially when working with neuropsychological phenomena. The mental and the physical are always running into each other. Is there an emergent something that is coordinating all of the brain's modules?

A CASE STUDY FROM
THE NEUROSURGICAL SUITE

The neurosurgeon Mark Rayport, who had practiced at the Medical College of Ohio at Toledo, made a stunning observation many years ago. During the course of craniotomies where the patient maintained consciousness, he would, unbeknownst to the patient, apply a small current of electrical stimulation to the olfactory bulb, the part of the brain that is heavily involved in managing the sense of smell. As Rayport tells the story, he would engage the patient in conversation with a positive tone, say, about the upcoming spring weekend. While chitchatting away, he would apply a pulse of electricity to the brain structure. The patient would suddenly interrupt the conversational flow and say something like, "Who brought the roses into the room?" Moments later, after Rayport had switched the conversation to negatively colored topics, he applied the same electrical impulse to the same exact place in the brain with the same exact intensity. The patient would again interrupt, but this time say, "Who brought the rotten eggs into the room?"[12]

Here was an example of a mental process constraining a brain process even though all of it was going on in the brain. It was as if a "top-down" mental process were informing a "bottom-up" physical biological process: the mind informing and influencing the brain. In short, while a mental state was generated by the physical brain, it too had a presence and could in turn influence the very physical state that produced it.

FLIRTING WITH EMERGENCE AND ITS IMPLICATIONS

Here is how emergence can be thought of. It occurs when a micro-level complex system organizes into a new structure, with new properties that previously did not exist, to form a new level of organization on the macro level.[13] For example, the behavior and properties of atoms are described by quantum mechanics. When those microscopic atoms come together to form a macroscopic baseball, however, a new set of behaviors and properties emerge that are then governed by Newton's Laws. Neither one predicts the other. Philip Anderson, a leading physicist at Princeton, wrote a famous article in the 1970s titled "More is different." In it he wrote that "the reductionist hypothesis does not by any means imply a 'constructionist' one: The ability to reduce everything to simple fundamental laws does not imply the ability to start from those laws and reconstruct the universe. In fact, the more the elementary particle physicists tell us about the nature of the fundamental laws, the less relevance they seem to have to the very real problems of the rest of science, much less to those of society."[14] That nailed it for me.

Nonetheless, the idea of emergence had a tough time being accepted, especially by neuroscientists. Why was it so tough? Hard reductionists have difficulty accepting that there is more than one level of organization—that different layers can contribute to the causal chain in understanding why things happen the way they do. Even if they do accept that, they can't accept the notion that the radical novelty that accompanies the emergence of a higher level cannot be predicted by lower-level events. Yet, multiple layers of organization are bread and butter to physicists, who faced these issues when quantum mechanics came on the scene. While some hard reductionists still lurk among the physicists, most believe that elements in nature are inherently unpredictable and, therefore, occur only with probabilities.

As I said, this is all very slippery and difficult to keep straight, and the arguments were energized fifty years earlier by Sperry. At a meeting at the Vatican he defiantly observed:

> This is not to say that in the practice of behavioral science we have to regard the brain as just a pawn of the physical and chemical forces that play in and around it. Far from it. Recall that a molecule in many respects is the master of its inner atoms and electrons. The latter are hauled and forced about in chemical interactions by the overall configurational properties of the whole molecule. At the same time, if our given molecule is itself part of a single-celled organism like paramecium, it in turn is obliged, with all its parts and its partners, to follow along a trail of events in time and space determined largely by the extrinsic overall dynamics of Paramecium caudatum. And similarly, when it comes to brains, remember always that the simpler electric, atomic, molecular, and cellular forces and laws, though

still present and operating, have all been superseded in brain dynamics by the configurational forces of higher level mechanisms. At the top, in the human brain, these include the powers of perception, cognition, memory, reason, judgment, and the like, the operational, causal effects of forces of which are equally or more potent in brain dynamics than are the outclassed inner chemical forces.[15]

DO YOU MEAN SUPERVENE OR SUPERSEDE?

Neuroscientists are so heavily reductionistic, as are most scientists, that Sperry's ideas really didn't take hold. In fact, Joe Bogen recounts in his lively autobiography how Sperry's colleagues at Caltech wanted him to get off the topic. At the same time, however, the ideas were being widely considered in the philosophical community and causing lots of thought and reaction. People were arguing about Sperry's use of *supersede* versus *supervene*. As the philosopher Sara Bernal points out, those of a "physicalist" stripe prefer the idea of "supervene" to "supersede."[16] Many years after Sperry's talk, the idea of supervenience was ably explained to me by Donald Davidson, the distinguished philosopher from the University of California, Berkeley. Davidson once attended a small meeting I held at the Hotel Bel-Air with George Miller, Leon, and others. His way of capturing the idea was to say, "supervenience might be taken to mean that there cannot be two events alike in all physical respects but differing in some mental respects, or that an object cannot alter in some mental respects without altering in some physical respects."[17] Others, including the philosopher David Lewis, have given the example of a dot-matrix picture: "A dot-matrix picture has global properties—it is symmetrical, it is cluttered, and whatnot—and yet all there is to the picture is dots and non-dots at each point of the matrix. The global properties are nothing but patterns in the dots. They supervene: no two pictures could differ in their global properties without differing, somewhere, in whether there is or there isn't a dot."[18]

So, the supervenience argument goes, no global, upper-level difference without local, lower-level differences. A supervenience physicalist/materialist holds that psychological, social, and biological levels supervene on physical and chemical levels. When Sperry says "supersede" and "outclassed," he vaguely suggests something other than supervenience—a picture where *level n* floats freer than *level n-1*. The unrepentant reductionists see a sleight of hand here and claim the determinist Sperry is suddenly talking about something else rather than neuronal cell firings.

Still, I am in the camp of Somerset Maugham, who famously observed that he had to be told stuff at least twice in order for it to stick. Someone once observed that a fanatic is someone who doesn't change his mind and doesn't change the subject. I am not a fanatic, but I remain unsettled on how all of those modules are organized and coordinated to give rise to unitary psychological experience. Is it enough to trot in the idea of emergence and claim victory? Getting a handle on the idea of emergence, and what it might or might not be, found me at the door of John Doyle, a mathematician at Caltech.

Doyle couldn't be more different than me. His constant thinking can only occasionally be interrupted by a martini, if force fed. He is also a jock. In the mid 1990s, he set, lost, set, and lost the rowing world record in the 40–45 age group, won a world championship in human-powered vehicles, won two golds (rowing), a fourth place (cycling) and a sixth place (triathlon) in the 1995 World Masters Games in Brisbane, Australia. Even though he is a highly sophisticated mathematician, he speaks plain English, a prerequisite for a conversation with me. To my great surprise, when I asked him one day how come he is so clear in his expositions, he matter-of-factly said, "Oh, I used to be an actor."

IT'S NEVER TOO LATE TO LEARN

Doyle is a professor of control and dynamical systems, a highly mathematical field full of difficult and challenging engineering problems that run from understanding turbulence to understanding the Internet. With his engineering background, Doyle thinks deeply about the architecture of systems. Any system. How are they organized to do what they do? Is there a universal architecture common to all information processing systems, such as brains, bacteria, cells, and corporate structures? Human-made stuff, of course, has a design and architecture to it. Maybe in the biologic world the forces of natural selection wound up producing entities that have a similar logic to their organization. Maybe if it is interacting parts that make for a global function, then all of these systems do have a similar architecture. At the heart of his search was a disbelief in the idea of emergence, which he finds spooky and undefined. From his engineering perspective Doyle was trying to understand the levels of explanation from the concrete perspective of actually designing and building something. When a something is actually built and carries out its function, it frequently seems as if it had emergent properties, but it doesn't. It should be understood in terms of its interacting parts.

Borrowing from the field of computer science, Doyle asks: What can we learn from the amazing systems we humans have built to process information, and how can we apply that knowledge to the problem of how the brain performs its tricks? It is commonplace in computer science to speak of a "layered architecture" of systems that build on one another, with one layer of function serving as the platform for the next layer of function. In the computer world they think in terms of seven layers. The top layer is the application or program being used, such as Facebook, while the bottom layer is the actual hardware such as an iPhone. Each layer, while inhabiting the other layers, is remarkably independent of them. Understanding this formulation is the trick. Can an engineering viewpoint help us think about a neurobiologist's problem? I think it can.

MIND/BRAIN NETWORKS, LAYERING, AND THE BRAIN

Layered architectures are one particular kind of modular architecture. Each layer can be thought of as a module. And, as I have said, there is lots of evidence that modular architecture is selected for in evolution and development because it allows one module to adapt through some sort of change without screwing up all the other modules. Layering, though, is a particular kind of modular architecture in that the layers (modules) are organized in a line. Layer 1 goes to Layer 2 goes to Layer 3, goes to Layer 4. It is not known if this is what the brain actually uses. Instead, it may use a hierarchical modularity consisting of many modules at each of the different scales (for example, neurons, circuits, lobes). Layering suggests a one-directional arrow (up or down through the layers) whereas hierarchical modularity enables a complex set of interactions between modules within a single scale or between disparate scales.

WHY LAYERING IS A HELPFUL CONCEPT

If you remove the cover and peek inside a mechanical clock, you will see a bunch of interconnected wheels, gears, and springs. There it is, churning away to produce a timekeeper. It doesn't know it's doing that, and the parts don't know anything about its function. Similarly in the brain, the individual neurons that churn away to produce our personal conscious experience know not what they do. In order to understand the various parts' mechanisms behind a simple clock, it quickly becomes clear that one needs to think in terms other than "this wheel connects to that spring and then to that wheel." The old "A connects to B which connects to C" story will get you nowhere.

Now think in terms of layers: five of them when it comes to clocks. Seeing the device in terms of layers, its architecture becomes evident, as does the way all mechanical clocks work. There is the energy layer, the distribution layer, the escapement layer, the controller layer, and the time indicator layer. First, a clock needs energy to make it work, so a spring needs to be wound up. That energy has to be stored and then slowly released. Second, wheels distribute that energy throughout the clock. Third, escapement mechanisms stop the energy from escaping all at once. Fourth, the controller mechanism controls the escapement function. Finally, all of this comes together to the fifth layer, which indicates the time. Notice, as you move up through the layers, each one does not predict the functional role of the next layer. The energy layer has nothing to do with the escapement layer, and so on.

Now note that each layer is flexible and largely independent. It is easy to swap in a new energy layer: The spring could be replaced by a weight and gravity, or possibly by batteries and motors, provided these are compatible with the core architecture. If, however, you changed to a new architecture, say, solid state electronics, most of the old parts would be obsolete. With the new architecture, you would still have a variety of swappable energy sources, including solar, but they are different from what can be swapped into a mechanical clock. The spring is gone, as are weight and gravity. Up at the time indicator layer, an infinite number of user interfaces indicating the time can be used, all of which are also independent and swappable: The new clock could even look like the old clock on the outside. So with layering, a great variety on the outside can hide a common core, or a common behavior can be implemented many different ways. As Doyle says, "Without layering you don't get this, or understand it." Again, without the organizing idea of layers, it would be extremely difficult to describe how a simple mechanical clock works or to build one. Over time, just as the clockmaker has figured out which parts work best, which size to use, which leveraging system to use, which wheels and springs, and so on, so natural selection did for our brains.

The tension at this point is that on the one hand, it seems that an abstraction is just another layer and thus a thing, a something you can hang your hat on. On the other hand is the view that an abstraction is not a mysterious thing, but a way of handling all the parts. Very recent eye-opening work by the neuroscientist Giulio Tononi and his colleagues has quantified how the layers might interact, and how the macro layers may indeed jump into the causal chain of command, just like Sperry suggested fifty years ago.[19] The battle for understanding the difference, if any, between "supersede" and "supervene" is on.

WINDING DOWN

As I have already said, fifty years ago, all that neuroscientists thought about were simple linear relationships—A makes B happen, and a thorough description of A is B. It was a reductionist heaven, and even today, that is how most neuroscientists view their work. This line of thinking left many of us banging our heads against the wall when we attempted to conceptualize how the mind is actually to be understood by our toils with the brain. We continued performing and interpreting linear experiments, putting off the bigger questions of how it all works together. Insights from people like Doyle, suggesting that we should think about the mind as interconnected networks of layers instead of linear relationships, are appreciated in some quarters. These insights, however, are hardly pervasive. Luckily, the general intellectual landscape has begun to change. The field of cell and molecular biology has come to realize that the object of its study is not to be understood in terms of working out linear pathways but instead by looking at the multiple interactions of a dynamic system.

On March 28, 2001, the cover of *Time* magazine showed a picture of the cancer drug Gleevec with the headline, "There is new ammunition in the war against Cancer. These are the bullets."[20] In 2001, most cancer biologists thought cancer was caused by a mutated

protein, and this mutated protein caused the cells to rapidly prolifer-
ate and avoid death. The simple thought was that if the protein could
be inhibited, then cancer would be eliminated. The drug Gleevec
inhibits a mutated protein (Bcr-Abl) that is found only in certain
types of chronic myeloid leukemia and in gastrointestinal stromal
tumors. In patients with these cancers, taking Gleevec inhibits the
mutated protein and cures the cancer. Unfortunately, these seem to
be the only two cancers that respond in this way.

Researchers rapidly started identifying other mutated proteins in
cancers and designing drugs to inhibit their activity. For example, in
many melanomas, a mutation in another gene (BRAF) causes rapid
proliferation of cells along with protection from cell death. A drug
was designed to inhibit BRAF activity. When this drug was given
to melanoma cells with the BRAF mutation, they started to die, but
then arose again only to grow rapidly after treatment. Researchers
soon figured out that BRAF works in a *network* to promote cell proliferation, not in a single linear pathway. When BRAF gets inhibited,
the *network* shifts, enabling another protein (CRAF) to promote proliferation.

As a consequence of these many findings, the new thinking by
cancer biologists is that there is not one single mutation that drives
cancer. Instead, a whole signaling network changes to drive the cancer. To kill cancers, the network must be targeted in multiple places.
Since 2006, the entire field of cell and molecular biology has now
realized that they are dealing with systems with feedback loops,
controls, compensatory networks, and all kinds of distant forces
impinging and involved with any one single function they might be
interested in. The complex architecture of a cell must indicate that
the complex architecture of the brain is at least as challenging and
may be very similar in some respects.

Let's look again at the fifty-year-old results of Case W.J., which
remain telling. Disconnection of one part of the brain from another
does prove that specific nerve pathways are important: Their duties
may range from signaling basic sensory and motor information all

the way up to complex informational exchanges between the two half brains that deal with such things as orthographic and phonological information. At another level, however, W.J. didn't seem a whit different from his preoperative state. He walked, he talked, he understood the world as usual, and he gave his big engaging smile right on cue. He also had those islands of specialized functions: Only his left hemisphere could manage language, and only his right hemisphere could grasp spatial relationships.

As the ensuing fifty years of research unfolded, these initial glimpses of human brain organization were deepened and put into a broader context. We now know just how specific the brain can be in its local processing. We know the brain is full of modules. In fact, a fundamental strategy of the brain is to reduce any new challenge to a module that can operate more or less automatically, outside the immediate mechanisms of cognitive control.

All of this, of course, brings us back to the question, How do all of those peripheralized modules interact to produce the glorious psychological unity we all enjoy? Are they massively and intricately exchanging a code of some kind or is it something else? Is it more like a society where all the citizens (modules) vote and out of that comes (emerges) democracy, which in turns constrains those that vote? Or, let's try a related metaphor, the orchestra.

In the spring of 2013, I was asked to deliver the keynote address at the annual meeting of the Association for Psychological Science in Washington, D.C. The meeting lasts a few days and is jam-packed with empirical studies on simple to complex animals, behaviors, brains, and societies. Four days of data, and most of it good. I decided to kick things off in my lecture with an "orchestra" metaphor, and, in doing so, a phrase popped into my head that I couldn't shake. I found myself telling the crowd, "The brain works more by local gossip than by central planning." In the world of tweets, only minutes later, I was stuck with it. Sheesh, now I had to explain myself. I felt like our split-brain patient J.W. must have felt. A behavior burped out of me from a largely silent processor that had been calculating away on life's events and then, suddenly, presented itself at the APS meeting. Great. Now I had to bring it into my cognitive flow, and my interpreter module had to explain it. So I did my best, and this is an approximation of what I said.

Think of all of those different musical instruments that have to be coordinated for an orchestra to produce music. The musicians all have a shared musical language and are all reading from the same script, but the conductor has to keep them queued up to do their thing at the right moment in time and at the right amplitude. At first glance, it appears that the individual players are not all connected directly, but they are connected through feedback loops via the conductor, a giant hub, who coordinates the overall timing. When all of this is done exquisitely, music is made to the delight of all. I once saw Skitch Henderson take over the baton from a less skilled conductor. Playing with the exact same musicians on the exact same piece, the house went from ho-hum to shaking with glory for hours. Each musical player has hard constraints on what he can do in time and space: He has to play the same song, at the same tempo, and on a specific instrument that has to be played with specific body parts. Being part of a symphony orchestra, however, requires coordination of all of the players, even though they do not seem to be directly communicating with each other. Yet, the coordination of all of those localized processors, locally doing their thing, seems to be the key. The conductor *appears* to do it in an orchestra; how does the brain do it?

Still, the orchestra metaphor plays along with a comfortable linear way of thinking, the notion that something is in charge, or, well, orchestrating all the parts. Something was missing from this analogy—something big. Then I saw a YouTube clip of Leonard Bernstein magnificently standing in front of an orchestra yet not overtly conducting. His hands were not moving at all; he simply reacted, giving positive feedback with his face as the musicians did their thing. The local processors, the modules, were keyed in and expressing themselves right on the money. Bernstein was there to enjoy it, revel in it, not direct it. What the heck? He wasn't controlling anything. The orchestra was working all by itself! Something had to be happening—what was it? It seemed as if something like local gossip must indeed be at work. The separate musicians were doing their thing like the parts in a mechanical watch, only local interactions and cueing were going on, too.

After the talk, a dinner had been arranged and during cocktails, the exceptionally talented Ted Abel, a molecular neurobiologist, remarked to me that he was a clarinetist and played in many an orchestra. He said, "You know, even though the conductor is up in front of you, the real action is how the players cue each other. With the clarinet, making a rightward swirl versus a leftward swirl as you're playing cues your colleagues on where you are going with the piece. It is like local gossip."

The metaphor snaps the mind to another view, that the old linear idea of flow of information in the brain may be wrongheaded. Is the brain really organized like the pony express, letters being passed from one outpost to another until somehow it all works? I don't think so. Sure, connections are important and are the heart of the split-brain story. Sure, specialized regions do specific things, which is the heart of modern brain imaging studies. Sure, individual variation of human capacity reflects variations in brain structure, functions, and experience. But how does it all work? What is the architecture of the system that allows it to do all of the wonderful things organisms like humans do on a second-by-second basis?

On a larger canvas, I see this as *the* question for mind/brain research. One problem with framing the question as "How does mental unity come out of a modular brain?" is that, currently, young graduate students of neuroscience are not commonly trained with the tools to understand such an architecture. It requires new skills and knowledge from a new array of experts who, in the main, are housed in engineering departments. Fortunately, some others see it this way, too, and after two and a half years of seemingly endless and pointless paperwork, a group of us have established a new graduate program at the University of California, Santa Barbara, aimed at bringing control and dynamical thinking to issues in neuroscience.

Nicholas Meyer, the great writer, director, and author of several *Star Trek* episodes, recently observed that Shakespeare never gave stage directions in his works. Johann Sebastian Bach didn't give musical directions, either. Two of the greatest artists in the history of the world were the original believers in the "less is more" principle. It used to be that the audience's job was to infer the meaning of a story to bring their minds into alignment. They were to abstract a work of art into their own narrative and participate in the artistry themselves. Meyer observed that that has all but fallen away in the modern narrative. Everyone expects to be told how stories come out, and nothing is left to infer.

In closing, I would like to point out that Darwin gave the world another brilliant play, the theory of evolution. Scientists have been studying this masterpiece for almost two hundred years, with regular offerings about how his description of natural selection actually works. Unlike Shakespeare and Bach and more like a scientist, he would have told us how it worked if he had known. But like a playwright, he didn't send us down a preconceived path he might have held. He left the issue open for the scientific community of the future to figure out. He cleverly set up the question by observing that small differences of form and function arise within any group of animals. With time, a small difference that conferred survival and reproductive advantage prevailed over those members of the species that didn't share the trait, and it became dominant.

But look around. Look at the vast amount of variation in the animal kingdom. How could that all actually occur? The first approximate answer came about fifty years ago when it was discovered that heritable variation does not occur without mutations to the DNA of an organism. That was a huge insight and, of course, was built on the long-established knowledge base about DNA, which first started in 1869. Yet can rare and random mutational events explain *all* of the variation we observe? It doesn't seem possible, and Darwin's puzzle has been hanging over the scientific community for dozens of years.

Two inventive biologists, Marc Kirschner, head of systems biology at Harvard, and John Gerhart, from the University of California, Berkeley, have tackled the problem head-on in their dazzling book, *The Plausibility of Life,* and have set a new stage for thinking about Darwin's dilemma and, with it, an architecture for biologic life. Building on advances in molecular genetics of the last thirty years, they argue that there is something called "facilitated variation." It goes like this: It is now known that there are "conserved core processes" that make and operate an animal. Kirschner and Gerhart say these processes "are pretty much the same whether we scrutinize a jellyfish or a human. . . . The components and genes are largely the same in all animals. Almost every exquisite innovation that one examines in animals, such as an eye, hand or beak, is developed and operated by various of these conserved core processes and components. . . . We suggest that it is the regulation of these (core) processes. Regulatory components determine the combinations and amounts of core processes to be used in all the special traits of the animal."[21]

This has all the ring of a layered architecture. Indeed, it is now known that gene expression is regulated by other genes: A gene codes a protein that regulates the expression of other genes. The takeaway idea here is that all of the variation we see in the natural world is the result of mutations that occur on a small number of *regulatory* genes, not on the thousands upon thousands of workhorse genes attending to the business of the body. The many fewer regulatory genes control the replication and activation and deactivation of the multitude of specific genes that do the work of the organism. Mutate a regulatory gene and there can be a huge effect. Consequently, the fact that mutations are rare is consistent, and a possible theory exists to explain why such mutations are so effective. The only way for Kirschner and Gerhart to have gotten to this incredible insight was to abandon simple linear thinking and to think about layered systems.

For students of mind and brain research, it's time to hitch up our pants, take a deep breath, and realize that much of the low-hanging fruit in neuroscience has been picked and packaged. The simple models have taken us only so far. My own view is that it is time to realize that the deep problems remain in full view and the answers are ripe for harvesting. Our job is to go after the deep problems with gusto and infer the answers from the underlying plots of the human play in front of us. It is a fabulous way to spend one's life.

EPILOGUE

MOST OF US CAN THINK BACK TO THE PEAK EXPERIENCES in our lives, most of which are highly personal. If life has been good to us, they are gratifying and ground our lives with personal meaning. For me, that afternoon at Caltech fifty years ago, when W.J.'s right brain completed an act of which his own left hemisphere had no knowledge, was, and continues to be, seared into my mind. I was stunned. This event landed me in a world of human inquiry that had an almost timeless origin that I certainly did not sense then. More than fifty years later, as I continue to try to understand the full meaning of that elementary and original finding, I do realize that I have only participated in this saga and still do not have its ending. No one does, and no one will for some time.

Gratifyingly, much has been learned from split-brain research. Starting with the original characterization that surgically disconnecting the two half brains resulted in someone with two minds, all the way to today's counterintuitive view that each of us actually has multiple minds that seem to be able to implement decisions into action, split-brain research has revealed and continues to reveal some of the brain's well-kept secrets. Nonetheless, what magical tricks the brain uses for taking a confederation of local processors and linking them to make what appears to be a unified mind, a mind with a personal psychological signature, is still a big unknown and the central question of neuroscience.

Discovering that a simple surgical intervention could produce two mental systems, each with its own sense of purpose and quite independent of the other, was the first shocker back in 1960. Gradually realizing that mind left and mind right were each aggregators of still other mental systems, dozens, if not thousands of them, focused our attention on how these systems interact. Do the separate systems have to be physically connected like bulbs on a string of Christmas lights or can they signal one another to act through other information channels? For instance, when a tree limb adds branches, the limb does not send a signal to the cells at the crotch of the tree to add more cells for support. The added physical weight of the new branches is detected locally by the cells at the crotch, and they automatically respond by making more cells to beef up the support. There is not a direct, privileged, discrete signal to add more cells. This process can only be understood by considering the whole physical reality of a tree. Likewise in the brain, there are many cueing systems other than neuron-to-neuron communication, from the ever-present oscillations of brain activity to local metabolic cueing systems. The interaction of discrete brain systems must involve all of these mechanisms and more.

Adding to our understanding of why the brain seems undisturbed by disconnections was not only the notion that it was, in a sense, sending half its decisions into the realm of the unconscious; it was also the discovery of the "interpreter." This special left brain system kept note of all the behaviors that resulted from the many mental systems. It appeared to be the surveillance camera on our behavior, which, of course, was the evidence that a mental or cognitive act had occurred. The interpreter not only took note; it tried to make "sense" out of the behavior by keeping a running narrative going on about why a string of behaviors was occurring. It is a precious device and most likely uniquely human. It is working in us all the time as we try to explain why we like something or have a particular opinion, or rationalize something we have done. It is the interpreter device that takes the inputs from the massively modularized and automatic brain of ours and creates order from chaos. It comes up with the "makes sense" explanation that leads us to believe in a certain form of essentialism, that is, that we are a unified conscious agent. Nice try, interpreter!

As I look back on my story, I realize I too have been conditioned by my trade to desire an ending, a summing up of my research. After sitting and listening to thousands of seminars over the years, I am all too familiar with the sentiment, "Does this guy know there is supposed to be a beginning, a middle, and an end?" The individual experimental science program is supposed to have such a structure, even though there are legions of scientists who don't seem to know how to present it that way. We live in the era of the "bottom line" mentality, with TED talks, sound bites, and news summaries. There is so much information to digest, we can only hope to grasp the world with compact and seemingly complete stories. We don't want to be left dangling.

We are all suckers for this information diet, and we all have come to depend on it, just like we have all succumbed to the instant gratification of texting and cell phones. And yet what separates the dilettante from the sophisticate is the appreciation that everything is not simple. The trick seems to be able to talk clearly while remaining fully aware of the underlying complexity of any story. For me it is the overwhelming realization that when trying to figure out how the brain does its masterful trick of enabling minds, we are barely at the starting line. Dig as deep as you want into human history: As long as there is a written record of thought, there is a record of humans wondering about the nature of life. It becomes obvious that all of us are just hopping into an ongoing conversation, not structuring one with a beginning, a middle, and an end. Humans may have discovered some of the constraints on the thought processes, but we have not yet been able to tell the full story.

ACKNOWLEDGMENTS

IRST AND FOREMOST, I WOULD LIKE TO CLOSE WITH A toast to all of our "split-brain" participants. Without their generosity, dedication, long hours, and endless patience we would never have learned as much as we have about the brain's structure and function. All of them worked hard, and all of us enjoyed our time together over these many years.

Second, the dozens of scientists that have participated in not only the studies reported in this book, but also the many other studies carried out over the past fifty years, deserve my deepest gratitude. Many were graduate students, postdoctoral students, faculty, and visitors from other institutions. All of them were as captivated as I was by the patients and their devotion to the research enterprise. They did great work.

In preparing this book I want to offer special thanks to several colleagues who read it in full and offered many helpful suggestions. I will list them alphabetically: Floyd Bloom, Leo Chalupa, Scott Grafton, Steven Hillyard, Michael Posner, Marc Raichle, and John Tooby. I also want to thank my wife, Charlotte; my sister Rebecca; and my good friends Dan Shapiro and Eric Kaplan—all offered extensive suggestions and edits. Finally, I could not carry out these assignments without the help of Jane Nevins at the Dana Foundation.

My steadfast agent, John Brockman, has always supported my efforts. He sticks with his people and keeps our eyes focused on the goals of scientific writing for the general public. For the past years, I have been fortunate to be with Dan Halpern at Ecco, HarperCollins. Dan spotted something in my little book on ethics and has been my publisher ever since. Thanks also go to Kallie Hill, my undergraduate research assistant who helped tremendously with the videos and the referencing. Lastly, my thanks to Hilary Redmon, my editor. She wrestled this wandering manuscript into coherence and always with a smile. I am in her debt.

1981 NOBEL PRIZE FOR PHYSIOLOGY OR MEDICINE[1]

Adapted from an article previously published in Science, October 30, 1981.

The 1981 Nobel Prize for Physiology or Medicine was awarded to three American-based scientists. Half of the prize went to Roger W. Sperry at the California Institute of Technology; the other half was awarded jointly to David H. Hubel and Torsten N. Wiesel of Harvard University.

Upon hearing the first news bulletin that Roger Wolcott Sperry, Ph.D., had been awarded the 1981 Nobel Prize for Physiology or Medicine, his colleagues and students could ask only the question, "Which aspect of his work was being rewarded?" Prior to actually knowing, there were at least three major areas of research that seemed deserving—developmental neurobiology, experimental psychobiology, and human split-brain studies. It was, of course, the final body of work that was honored, but disciples of the other studies remain convinced that the other approaches were just as deserving.

The Nobel award to Sperry, professor of psychobiology in the Division of Biology at the California Institute of Technology, serves as an inspiration to those who believe that understanding the human conscious process is the ultimate objective of neuroscience and that

it can be studied with scientific rigor. It represents a grand appreciation of Roger W. Sperry for his relentless pursuit of an understanding
of the conscious processes of the human brain, a pursuit he began
with related but more fundamental studies more than forty years ago
and maintained with a singular excellence and passionate energy. In
fact, it can be said that it is Roger Sperry's overall body of work that
has served to conceptualize the objectives and questions pursued in
much of current neuroscience.

The particular studies of the human brain cited in the Nobel
award began in the early 1960s, and the application of the initial
insight gained from these split-brain studies to subsequent brain
research has all the earmarks of a Sperry enterprise. It all started
in 1961 when Joseph E. Bogen, M.D., proposed split-brain surgery
be carried out on a forty-eight-year-old war veteran in an effort to
control otherwise intractable epilepsy. Bogen was aware of Sperry's
earlier work on severing the connections between the hemispheres in
animals, and Sperry and Ronald E. Myers had already demonstrated
striking disconnection effects, that is, that information learned by
one half brain did not transfer to the other. At the time of the human
studies, the animal paradigm was already in pervasive use in experimental laboratories around the world.

In fact, the animal work done by Sperry stood in dramatic contrast to prior human work on callosum-sectioned patients that had
been carried out in the early 1940s. These early reports suggested
that cutting the forebrain commissures, as they are called, had no
detectable effect on interhemispheric communication. It was these
studies, in part, that discouraged the view that discrete pathways
in the brain carried specific kinds of information. There was some
question about the usefulness of the surgical technique as well for
controlling epilepsy, but Bogen, after a careful review of the medical cases, concluded there was a good chance the surgery should
help. That proved correct. In this new light, the stage was also set for
new experimental observations on split-brain humans—a task made

possible over the years by the generous cooperation of the patients themselves.

No one was prepared for the riveting experience of observing a split-brain patient generating integrated activities with the mute right hemisphere that the language-dominant left hemisphere was unable to describe or comprehend. That was the sweetest afternoon. It was clear that the animal model held for humans, and, as a result, Sperry masterminded a program of human split-brain research that continues today. The implications of these findings for theories of consciousness and cerebral specialization, for cognitive science and clinical neurology, and even for thoughts about human values were all developed in Sperry's laboratory. He was exceedingly generous to a series of students who went through Caltech, including Colwyn Trevarthen, Jerre Levy, Robert Nebes, Charles Hamilton, Eran Zaidel, and myself, all of whom assisted in developing the split-brain story. Yet the overall achievement was Roger Sperry's. He is constitutionally able to be interested in only critical issues, and he drove this herd of young scientists to consider nothing but the big questions.

There were two main phases of the human work in Sperry's laboratory. The first was to characterize the basic neurologic and psychologic consequences of split-brain surgery and to identify the individual psychological nature of each separated hemisphere. Results accumulated over a period of six years demonstrated that the cortical commissures were critical to the interhemispheric integration of perceptual and motor functions. These studies also revealed that the mute right hemisphere was specialized for certain functions that dealt with nonverbal processes, while, not surprisingly, the left hemisphere was dominant for language. For the first time in the history of brain science, the specialized functions of each hemisphere could be positively demonstrated as a function of which hemisphere was asked to respond. The important clinical observation of brain-damaged patients had only been able to show absence of function— not the concurrent but separate and lateralized coexistence of such functions. Finally, the implications for a theory of mind were abundantly clear after observing the patient's lack of awareness in one half brain about the activities of the other.

The second phase of study emphasized the different cognitive styles of the hemispheres and the special linguistic capacities of the right half brain. These findings were pursued, not only by Sperry but also by other researchers investigating the lateralization story, and have included observations of both neurologically damaged and normal populations. All of this has given rise to a wealth of possibilities concerning the nature of human brain organization. The issues raised are of great interest, and pursuit of hard answers to questions central to this work comprises much of the contemporary research in neuropsychology.

It has to be kept in mind that this body of work was preceded by a series of studies by Roger Sperry that laid the groundwork for much of the present-day field of developmental neurobiology—the experiments of which probably consume about half of all activities of neuroscientists. It all began at the University of Chicago in the 1940s. Graduate student Sperry challenged the neurobiologic theory of his brilliant mentor, Paul Weiss, that "function precedes form," that is, that the central nervous system and its peripheral connections were not specified by genetic mechanisms. In a series of experiments that extended over twenty years, each more spectacular than the last, Sperry developed his theory of chemospecificity. His conception that chemical gradients are critical to the specification of cell-to-cell connections is still at the center of current neurobiological work, and every modern-day developmental neuroscientist is trying to find the loophole.

After Chicago, Sperry went to Yerkes Laboratory and spent some important time with Karl Lashley. Once again Sperry intuitively rejected the going model of cerebral function and challenged Lashley's theories on equipotentiality and mass action. While carrying out new studies, which to some extent led to the animal discoveries in split-brain work, he also put to rest a few theories Gestalt psychologists had about brain mechanisms and perceptual processes. In the early 1950s, Sperry, already recognized as a world leader in brain research, was invited to be the Hixon Professor of Psychobiology at Caltech by Nobel Laureate George W. Beadle. It was a prime job in a glorious institution, and Sperry settled in and started his major systematic work with both animals and humans in split-brain research.

Life in science today is not as much fun as it used to be. It is full of time-consuming, boring administrative chores, of bureaucratic double talk, of responding to endless mediocre demands for "programmatic applications" in science pursuits and grant writing, and all the rest. As the dollars allocated for science decrease in number, as they have done for the last fifteen years, the request for articulated trivia goes up, and some people are actually beginning to think that this is science. We all know this, and every time I have to deal with it, I think of Sperry. He was unable to be scientifically trivial. He scowled when people proposed an extensive series of experiments. He knew how science really works, how things just happen, and that leads then are actively pursued, and pursued with vigor. He never played the bureaucratic game; he never gave in to the forces of trivia, and I hope his steadfast ways with their now grand rewards will signal the larger community to set things straight once again. Those were happy days working in his lab, trying to keep up with the intellectual excitement and freedom he always so brilliantly engendered.

Sperry's dazzling career had its origins in a time when brain scientists, then not so chic, studied the brain because they were interested in how its workings explained behavior. In some sense they were not interested in the brain per se, as are so many current-day neuroscientists. Their experiments constantly focused on discerning something about how the biologic system worked to support behavior, and ultimately the generation of conscious awareness. Roger W. Sperry, even while studying individual neurospecificity, saw and talked about its implications for the broader problems of nature versus nurture, a theme also so eloquently investigated by fellow award winners David Hubel and Torsten Wiesel.

Another example of his functional approach was Sperry's brilliant paper on how certain aspects of fish behavior changed after selective surgical manipulation, which generated an "efferent copy theory," a theory that is central in most perceptual-motor research today. There were also those classic theoretical papers of the 1950s on "the neural basis of the conditioned response" and "neurology and the mind-brain problem." In short, Roger Sperry was a neuroscientist who was perfectly clear about why he chose to study the brain. He worked to help elucidate the biological and psychological nature of man, a problem by no means solved, but a problem he helped define and advance knowledge about like no other scientist in the history of the world.

THE EDITOR OF *SCIENCE* magazine and Roger Sperry and Joe Bogen were all pleased and generously responded to this *Science* article with exceedingly kind notes:

October 21, 1981
From *Science*:
Dear Dr. Gazzaniga:

Your sketch describing the contributions and experimental approach of Professor Sperry was elegant and informative. The descriptions of the creative atmosphere that pervaded the laboratory will evoke resonance in others who have been privileged to feel the excitement of science at the frontier. I am sure that our elite readership will sense the magic that you have sought to convey. We very much appreciate your willingness to provide the material quickly. . . .

Yours sincerely
Philip H. Abelson
Editor

October 29, 1981
California Institute of Technology
Dear Mike:

Have just read your article in *Science* and hasten to extend my deepest gratitude. You rose magnificently to the occasion overriding our personal differences with a statement that I hope and believe will always stand as a lasting credit to yourself and the rest of us involved. Of course, I feel you exaggerated my role in the split-brain developments but trust this is something most readers will quickly sense anyhow. Again, I owe you.

Thanks also for your very nice wire and all the best.

Sincerely,

Roger

October 30, 1981
New Hope Pain Center
(handwritten)
Dear Mike:

I wanted to write to express my appreciation for your appreciation (in Science 30 Oct p 517) of Roger, not only for your generous references to me and to others, but also for the super way in which you used this opportunity to make some very important points.

Altho' Roger may not say much (has he ever?), I've no doubt that he wanted very much to have said the things you said so well.

Joe

I asked George Miller, "Just what is it that cognitive science wants to know?" The following week, the guiding ideas behind cognitive neuroscience took form in a long memo he wrote to me, which I present in edited form:

To: Michael S. Gazzaniga
From: George A. Miller
Re: "COGNITIVE SCIENCE"

An intense undergraduate, in the sharp panic of an identity crisis, rushed to his professor: "I don't know who I am. Tell me, who am I?" The professor replied wearily, "Please, who's asking the question?"

The story flashed to mind recently when a friend, who watches science with the eyes of a biologist, asked: "What do cognitive scientists want to know?" Anyone capable of posing such a question must already know the answer. To know is to have direct cognition of. Obviously, scientists of cognition want to have direct cognition of having direct cognition. Any etymologist could tell you that.

What would a biologist accept as an answer? Something deep is called for. My friend is not asking about computers, or simulations, or logical formalisms, or the latest methods of psychological experimentation—none of that ancillary horseshit that fills so much of the conversation of cognitive scientists. A deeper answer is that cognitive scientists want to know the cognitive rules that people follow and the knowledge representations that those rules operate on. But this language—cognitive rules, knowledge representations—is precisely the kind of smoke that started my friend looking for a fire.

Let us begin with a question that we can answer: What do biologists want to know? Biologists want to discover the molecular logic of the living state. What is the molecular logic of the living state? Simple. It is the set of principles that, in addition to the principles of physics and chemistry, operate to govern the behavior of inanimate matter in living systems. (That is an almost direct quotation from the introduction to a biochemistry textbook.)

Is this the kind of answer a biologist expects when he asks what cognitive scientists want to know? If so, perhaps we can construct an answer based on this model of what an answer should be. Because I am a little slow at these games, however, I shall take three steps to get where I am going. First, I will substitute psychologists for biologists. No substitution seems required for molecular logic; I assume that "molecular" in this context means "susceptible to analysis," and is not limited to the analysis of matter into chemical molecules. And then I will substitute conscious for living, because I consider consciousness to be the constitutive problem for psychology, just as life is the constitutive problem for biology. Now I have achieved the following: Psychologists want to discover the molecular logic of the conscious state. So far so good. But now what do we mean by molecular logic of the conscious state? Let's see if substitution leads anywhere: the set of principles that, in addition to the principles of physics, chemistry, and biology, operate to govern the behavior of inanimate matter in conscious systems. These substitutions say little more than that psychology is the next step in the positivistic hierarchy of sciences. The result sounds pretty good to me, but can I follow through? That is to say, the biochemist whose formulation I have borrowed as my model had a large and impressive textbook full of biological principles to illustrate what he was talking about. What do I have?

One thing I do not have is behaviorism, because most behaviorists are dedicated to the proposition that consciousness is irrelevant to the science of psychology. Another thing I do not have is artificial intelligence, because computer simulations have no need for the psychological distinction between living and nonliving systems, for that matter.

What I seem to have is a way of looking at psychology, a criterion to keep in mind while thumbing through psychological handbooks. It might be formulated like this: Any behavior that is unaffected by the state of consciousness of the behaving system is of no concern to psychology. Dreaming, for example, is a concern of psychology, because if you wake up if your state of consciousness changes— dreaming is affected. . . . The ability to violate some principle by an act of will is now the critical test that the principle in question is one that is relevant to psychology. . . . The problem, however, is that my friend did not ask what psychologists want to know. He asked what cognitive scientists want to know.

A second set of substitutions can be tried, therefore. Suppose we substitute states of knowledge for the conscious state. Then we obtain: Cognitive psychologists want to discover the molecular logic of states of knowledge, where the molecular logic of states of knowledge refers to the set of principles that, in addition to the principles of physics and chemistry, govern the behavior of inanimate matter in knowledge systems. Reference to biological and psychological principles is here omitted, for the computers can instantiate knowledge systems; computers need obey no biological or psychological principles.

The criterion for looking at research would now become: Any behavior that is unaffected by the state of knowledge of the behaving system is of no concern to cognitive science. If you turn off the power in a computer, for example, the consequences will not depend on the state of knowledge of the computer, so they would be of no concern to cognitive scientists. . . .

I have no desire to dissuade anyone who wants to develop cognitive science along these lines, but neither do I have any desire to join with them. I would prefer to take a different line, defining still another science more narrowly. So I will now take a third step, as follows: Cognitive neuroscientists want to discover the molecular logic of epistemic systems, where the molecular logic in question this time is the principle that, in addition to the principles of physics, chemistry, biology, and psychology, governs the behavior of inanimate matter in epistemic systems. (The term "epistemic system" is negotiable; I use it as a placeholder for something better.) A further substitution is possible: animate for inanimate in the final clause. I am unclear whether it would make any real difference.

By including the requirement that cognitive neuroscience is concerned only with living, conscious systems, we cut artificial intelligence free to develop in its own way, independent of the solutions that organic evolution happens to have produced. Now our concern is for a subset of conscious systems, and the criterion is whether or not the system's state of knowledge affects its behavior. . . .

It should be clear by now that I really don't have an answer to the question, what do cognitive scientists want to know? But I think that cognitive neuroscientists want to know something that is reasonably interesting, and that there really might be some promise in following up systematically the implications of the definitions that we arrived at by substitution into our biological model.

Unbelievable as it may seem, I attempted a response. After all, it was spring.

To: George A. Miller
From: Michael S. Gazzaniga
Re: Exemplars of Cognitive Neuroscience

O.K., your claim is that our task is to understand those processes active in living systems that can exert control over the comings and goings of a variety of mental constituents that make up a cognitive agent. (Put differently, is it also fair to say that the defining qualities of a cognitive system are coincident with an information processing disorder?) Alternatively, it is our task to understand cerebral software, the programming stuff that orchestrates the spatial-temporal patterns of the neural network. First, has your definition of cognitive neuroscience moved the ball down the field? I think it has. Consider what others have said about what cognition is, usually using other terminologies. Sperry, for example, used to argue that consciousness is an emergent property of the spatial-temporal interaction of the neuronal system subserving the phenomenon. He maintained that these emergent mental properties feed back, as it were, and control the activities of the system that produced it. To me this position is a neuroscientist's way of saying *cognitive act*. MacKay's hypothesis on what the cardinal feature of a cognitive system is goes like this: "the direct correlate of conscious experience is the self-evaluating, supervisory or metaorganizing activity of the cerebral system and it is this system that determines norms and priorities and organizes the internal state of readiness to reckon with the sources of sensory stimulation." That strikes me as a rather passive description of the conscious process and it takes on more of the character of a "jobber" or "dispatcher." He does not characterize the system as one that tries to penetrate the organism's natural tendency to reflexively respond to a command.

If I am right, your definition has advanced at least my understanding of some issues and has clearly stated that the task is to discover the rules that govern the epistemic system—the one living system that governs the biologic system. When thinking about that, I am maintaining that the epistemic system is supraordinate to the biologic system. Is that what you were driving at?

At any rate, you have set us to the task of actually trying to figure out the principles of not only how cognitive systems announce their products to consciousness, but also the criterion that a cognitive system is a process that can supersede the cerebral architecture. How else can we illuminate this dynamic other than by studying disruptive brain states? In some sense the cognitive neuroscientist is trying to trick out of the organism insight into that puzzling problem. But before raising some problems from studies on brain-damaged patients, let me make one other observation that I think needs up-front analysis.

The kind of analysis one would bring to understanding a New Yorker as opposed to understanding New York would be quite different. The kind of analysis one brings to understanding a serial system as opposed to a parallel system also seems to me to be quite different. Before we proceed with an intelligent analysis of cognitive function, do we have to face up to the issue as to whether or not the system is in fact competing for the attention of the person? If we agree that this is a reasonable model, crudely put at this point, then it seems to me how one approaches problems in brain disease that merit consideration for a theory of cognition becomes quite different.

Let me now consider a brain disease situation that speaks to this notion of what constitutes a cognitive system. There can be in brain disease relatively discrete disruptions of one of the system properties of the cognitive agent. It is common, for example, to study patients with memory dysfunctions. On one level of analysis they are unable to (1) retain new information and (2) combine two new elements into a fresh concept. Looking into the pathophysiology underlying these disorders, one finds that both diffuse and focal disease states correlate in this psychologic disarray. It is only on deeper probe that one begins to see differences at the psychologic level. Patients with focal disease possess a dense inability to transfer information from short-term to long-term memory, although lavishly assisted in their recall performance by cueing (e.g., categorical headings embedded in a long word list). On the other hand, patients with diffuse disease are not assisted by this cognitive strategy. Their recall performance stays down on the floor.

What are we to do with these observations? First of all, are we to dismiss the diffuse disease patients as still embodying a cognitive system? Has their agency been lost? If not, what is it about them that characterizes them as a member of this species? I don't have an answer. It seems to me that brain-diseased patients tell us immediately that we must bring more specificity to the definition of "cognitive penetrability" as a criterion for a cognitive system. I have the strong feeling that there is a real insight here, but a nagging feeling that we can too easily dismiss a lot of cognitive agents.

To which George replied:

To: Michael S. Gazzaniga
From: George A. Miller
Re: There's a long, long trail a'winding

Since you accept, at least tentatively, my definition of cognitive neuroscience, our next task is to try to put it to work. I want to restate the definition, but first I want to get rid of "epistemic system." Let me begin by pointing in the general direction I had in mind.

Organic Knowledge Systems. A "knowledge base" is any tangible collection of signals that are arranged according to some accepted coding scheme in order to represent a given body of information. A knowledge base coupled with an information processing system for using it (for storing, retrieving, erasing, comparing, searching, etc.) is a "knowledge system." Obviously, a knowledge base is useless except as part of a knowledge system that (unlike libraries or computers) is governed by biological and psychological principles, i.e., a living, animate, agentive knowledge system.

Definition of Cognitive Neuroscience. Cognitive neuroscientists attempt to discover the molecular logic of organic knowledge systems, i.e., the principles that, in addition to the principles of physics, chemistry, biology, and psychology, govern the behavior of inanimate matter in living knowledge systems.

The Cognitive Criterion. It follows from this definition that any behavior unaffected by the state of knowledge of the behaving system is of no concern to cognitive neuroscience.

Implications of Definition. This definition is compatible with various approaches to cognitive neuroscience: (1) Evolution of knowledge systems. For example, the evolutionary shift from genetically stored knowledge to knowledge acquired from experience. (2) Ontogenesis of knowledge systems. For example, the neural basis of personal memory. (3) Psychology of knowledge systems. For example, the effects of attention, as indicated by evoked potentials, perhaps, on knowledge-governed behavior. (4) Neurology of knowledge systems. For example, the correlation of different types of brain disease. And so on. None of these approaches is novel—which means that we could have something to say about each of them.

A philosophical objection to this approach is that, by introducing successive definitions of biology, psychology, and cognitive neuroscience in this manner, we have made it reductionistic. That is to say, the principles sought by the cognitive neuroscientist are also principles of psychology, and the principles sought by psychologists are also principles of biology. Since I have always thought of scientific psychology as a branch of biology, this objection carries little weight with me. It would carry greater weight, however, with such distinguished scientists as B. F. Skinner or H. A. Simon.

Implications of Criterion. A central question in your memo of June 1 might be phrased as follows: What are the operational implications of the claim that "any behavior unaffected by the state of the behaving system is of no concern to cognitive neuroscience"?

Several things occur to me when you press this button. First, Zenon Pylyshyn should not have to assume responsibility for this phrasing of the criterion. As I understand his notion of "cognitive penetrability," it is intended to discriminate between the fixed "architecture" and the modifiable programs for a mental computer. We, on the other hand, are trying to distinguish what cognitive neuroscientists want to know from what they leave to others. It is not clear to me, in my ignorance of Pylyshyn's ideas, whether these two distinctions coincide, so the only line I can try to develop is our own.

Second, I see two obvious ways to apply the criterion: (1) Change an organism's state of knowledge and try to demonstrate a resultant change in its thinking or behaving. Or (2) leave the organism's knowledge alone, but vary the materials used in a task to see whether thought or behavior changes as a function of their familiarity.

If I have understood your example, the case of a patient with diffuse brain disease illustrates one of the difficulties of applying the criterion in manner (1); since it is apparently impossible to change such a patient's state of knowledge, his memory-governed behavior was of no concern to cognitive neuroscience. For such a patient, therefore, it would be necessary to apply the criterion in manner (2)—essentially, to change the contents of the questions asked until we find something the patient does remember. Does this answer the disturbing question raised at the close of your memo?

Third, I would think of this criterion as something to guide us, as authors, in picking and choosing what studies to write about and how to organize them. I see nothing wrong with confessing that this is the criterion we used (if, indeed, we did), but it does not seem to me to be something that we must rub the reader's nose in.

Levels of Description. One of the biggest problems I have in trying to get my thoughts straight about cognitive neuroscience is that different people work at different levels of description, and no one pays attention to how his level is related to descriptions at other levels. I assume this degree of incoherence is possible because the different levels are only loosely related, which, if true, is an interesting observation in its own right.

The closest discussions I have seen of the level problem have come from the MIT Artificial Intelligence Laboratory, where I assume that Minsky and Marr have been the guiding lights. It is forced on anyone who works with computers, I guess. For example, in P. H. Winston's *Artificial Intelligence* (Addison-Wesley, 1977) eight levels of description of the operation of a computer are distinguished: (1) transistors, (2) flip flops and gates, (3) registers and data paths, (4) machine instructions, (5) compiler or interpreter, (6) LISP, (7) embedded pattern matcher, and (8) intelligent programs. D. Marr and T. Poggio (A theory of human stereo vision, *Proc. Royal Soc. London,* 1977) bring this closer to neurology when they distinguish four levels of description that should apply both to computers and to brains: (1) transistors and diodes, or neurons and synapses, (2) assemblies made from elements at level (1), e.g., memories, adders, multipliers, (3) the algorithm, or scheme for computation, and (4) the theory of the computation.

Clearly, most neuroscientists today are gung ho for level (1); neurotransmitters are hot stuff. I have also encountered a little work at level (2)—e.g., Mountcastle's description of columnar assemblies—so I assume there is more that I don't know about. Level (3) is as abstract as any neuroscientist had dared to dream about—maybe it has been achieved in such cases as Vince Dethier's analysis of flies. Level (4) has been neglected, and Marr and Poggio propose that it is the responsibility of artificial intelligence to provide general theories by which the necessary structure of computation at level (3) can be defined.

I hold no brief for either of these analyses, but I do agree with them that anything as complicated as a nervous system can be understood at several levels. And the logic of levels is such that they must be only loosely connected to one another—otherwise they would not be distinct levels. Moreover, the processes described at level N could probably be achieved by many higher processes at level N + 1—so a description at level N is never really an explanation of what is really going on at level N.

Problem. What do levels have to do with our definition of cognitive neuroscience? This is not a rhetorical question—I really need an answer.

For example, a particular drug known to affect synapses

in a given way (manipulation at level 1) is observed to affect behavior governed by the patient's general knowledge of spatial relations (a consequence at level 4). It meets our criterion (applied in manner 2) for inclusion in cognitive neuroscience. But to include it is not to understand it! Help!

NOTES

CHAPTER 1: DIVING INTO SCIENCE

1. R. Sperry, "The growth of nerve circuits," *Scientific American* 201 (1959): 68–75.
2. Told to me by Berkeley physics professor and Alvarez's former colleague Rich Muller.
3. Many of these biographical details have been reported in other recent academic reviews: M. S. Gazzaniga, autobiographical essay in L. R. Squire, ed., *The History of Neuroscience in Autobiography,* vol. 7 (New York: Oxford: Oxford University Press, 2011); M. S. Gazzaniga, "Shifting gears: Seeking new approaches for mind/brain mechanisms," *Annual Review of Psychology* 64 (2013): 1–20.
4. A. P. Aristides, "Spreading depression of activity in the cerebral cortex," *Journal of Neurophysiology* 7 (1944): 359–90.
5. Dr. Linus Pauling in conversation with me.
6. Variously attributed to Francis Bacon or Roger Bacon (see discussion: *Horse Teeth* at http://www.lhup.edu/~dsimanek/horse.htm).
7. K. S. Lashley, *Brain Mechanisms and Intelligence* (Chicago: University of Chicago Press, 1929).
8. R. W. Sperry, "Orderly functions with disordered structure," in H. V. Foerster and G. W. Zopt, eds., *Principles of Self-Organization* (New York: Pergamon Press, 1962), pp. 279–90.
9. D. Helfman, "Dr. Mead Livens Lounge," *California Tech* 62, no. 24 (1961): 1.
10. D. G. Attardi and R. W. Sperry, "Preferential selection of central pathways by regenerating optic fibers," *Neurology* 7 (1963): 46–64.
11. Dr. Mitch Glickstein, personal communication.
12. Dr. Roger Sperry in conversation with me.

13. Dr. Mitch Glickstein, personal communication.

14. Steve Allen et al., *Dialogues in Americanism* (Chicago: Henry Regnery, 1964).

CHAPTER 2: DISCOVERING A MIND DIVIDED

1. J. Bogen, autobiographical essay in L. R. Squire, ed., *The History of Neuroscience in Autobiography*, vol. 5 (San Diego: Elsevier Academic Press, 2006), p. 90.

2. J. D. Watson and F. H. Crick, "Molecular structure of nucleic acids; a structure for deoxyribose nucleic acid," *Nature* 171, no. 4356 (1953): 737–38.

3. M. S. Gazzaniga, J. E. Bogen, and R. W. Sperry, "Some functional effects of sectioning the cerebral commissures in man," *Proceedings of the National Academy of Science* 48 (1962): 1765–69; M. S. Gazzaniga, J. E. Bogen, and R. W. Sperry, "Laterality effects in somesthesis following cerebral commissurotomy in man," *Neuropsychologia* 1 (1963): 209–15; M. S. Gazzaniga, J. E. Bogen, and R. W. Sperry, "Observations on visual perception after disconnection of the cerebral hemispheres in man," *Brain* 88 (1965): 221–36; M. S. Gazzaniga, J. E. Bogen, and R. W. Sperry, "Dyspraxia following division of the cerebral commissures," *Archives of Neurology* 16 (1967): 606–12; M. S. Gazzaniga and R. W. Sperry, "Language after section of the cerebral commissures," *Brain* 90 (1967): 131–48.

4. R. E. Myers, "Interocular transfer of pattern discrimination in cats following section of crossed optic fibers," *Journal of Comparative & Physiological Psychology* 48, no. 6 (1955): 470–73.

5. R. E. Myers and R. W. Sperry, "Interocular transfer of a visual form discrimination habit in cats after section of the optic chiasm and corpus callosum," *Anatomical Record* 115 (1953): 351–52.

6. C. Morgan, *Physiological Psychology* (New York: McGraw-Hill, 1943).

7. C. Morgan and E. Stellar, *Physiological Psychology*, 2nd ed. (New York: McGraw-Hill, 1943).

8. P. Black and R. E. Myers, "Visual function of the forebrain commissures in the chimpanzee," *Science* 146, no. 3645 (1964): 799–800.

9. R. W. Sperry, "Mechanisms of neural maturation," in S. S. Stevens, ed., *Handbook of Experimental Psychology* (New York: Wiley, 1951).

10. R. W. Sperry, N. Miner, and R. E. Myers, "Visual pattern perception following subpial slicing and tantalum wire implantations in the visual cortex," *Journal of Comparative Physiological Psychology* 48 (1955): 50–58.

11. M. S. Gazzaniga, J. E. Bogen, and R. W. Sperry, R.W. (1962). "Some functional effects of sectioning the cerebral commissures in man," *Proceedings of the National Academy of Science* 48 (1962): 1765–69.

12. N. Geschwind and E. Kaplan, "A human cerebral deconnection syndrome: A preliminary report," *Neurology* 12 (1962): 675–85.

13. A. Damasio, "Norman Geschwind (1926–1984)," *Trends in Neuroscience* 8 (1985): 388–91.

14. N. Geschwind and E. Kaplan, "Human split-brain syndromes," *New England Journal of Medicine* 266 (1962): 1013.

15. B. Grafstein, autobiographical essay in Larry Squire, ed., *The History of Neuroscience in Autobiography*, vol. 3 (Oxford: Oxford University Press, 2001).

16. N. Geschwind, "Disconnexion syndromes in animals and man," *Brain* 88 (1965): 237–94.

17. J. Bogen, autobiographical essay, p. 87.

18. J. Rose and V. Mountcastle, "Touch and kinesthesis," in J. Field, ed., *Handbook of Physiology, Section 1: Neurophysiology* (Washington, D.C.: American Psychological Society, 1959), pp. 387–429.

19. M. S. Gazzaniga, J. E. Bogen, and R. W. Sperry, "Laterality effects in somesthesis following cerebral commissurotomy in man," *Neuropsychologia* 1 (1963): 209–15.

20. Bogen, autobiographical essay, p. 95.

21. O. Devinsky, "Norman Geschwind: Influence on his career and comments on his course on the neurology of behavior," *Epilepsy and Behavior* 15, no. 4 (2009): 413–16.

22. N. Wade, "American and Briton win Nobel for using chemists' test for M.R.I.'s," *New York Times,* Oct. 7, 2003.

23. J. Bogen, autobiographical essay.

24. C. B. Trevarthen, "Two mechanisms of vision in primates," *Psychologische Forschung* 31 (1968): 299–337.

25. M. S. Gazzaniga, "Cross-cueing mechanisms and ipsilateral eye-hand control in split-brain monkeys," *Experimental Neurology* 23 (1969): 11–17.

26. J. E. Bogen and M. S. Gazzaniga, "Cerebral commissurotomy in man: Minor hemisphere dominance for certain visuospatial functions," *Journal of Neurosurgery* 23 (1965): 394–99.

27. M. S. Gazzaniga, "Effects of commissurotomy on a preoperatively learned visual discrimination," *Experimental Neurology* 8 (1963): 14–19.

CHAPTER 3:
SEARCHING FOR THE BRAIN'S MORSE CODE

1. M. S. Gazzaniga, "Interhemispheric cueing systems remaining after section of neocortical commissures in monkeys," *Experimental Neurology* 16 (1966): 28–35.

2. M. S. Gazzaniga and S. Hillyard, "Language and speech capacity of the right hemisphere," *Neuropsychologia* 9 (1971): 273–80.

3. L.B., personal communication

4. M. S. Gazzaniga, J. E. Bogen, and R. W. Sperry, "Observations on visual perception after disconnection of the cerebral hemispheres in man," *Brain* 88 (1965): 221–36.

5. M. S. Gazzaniga and R. W. Sperry, "Language after section of the cerebral commissures," *Brain* 90 (1967): 131–48.

6. M. M. Steriade and R. W. McCarley, *Brain Control of Wakefulness and Sleep,* 2nd ed. (New York: Plenum, 2005).

7. G. Berlucchi, M. S. Gazzzaniga, and G. Rizzolatti, "Microelectrode analysis of transfer of visual information by the corpus callosum," *Archives Italiennes de Biologie* 105 (1967): 583–96.

8. D. Hubel, *David (1995) Eye, Brain, Vision* (New York: Scientific American Library, 1995). Series (Book 22).

9. R. A. Filbey and M. S. Gazzaniga, "Splitting the brain with reaction time," *Psychonomic Science* 17 (1969): 335–36.

10. See G. Berlucchi, "Visual interhemispheric communication and callosal connections of the occipital lobes," *Cortex* (2013); S0010-9452(13)00037-3; doi: 10.1016/j.cortex.2013.02.001.

11. D. Premack, "Reversibility of reinforcement relation," *Science* 136, no. 3512 (1962): 255–57.

12. C. Blakemore and D. E. Mitchell, "Environmental modification of the visual cortex and the neural basis of learning and memory," *Nature* 241 (1973): 467–68.

13. M. S. Gazzaniga, "Cross-cueing mechanisms and ipsilateral eye-hand control in split-brain monkeys," *Experimental Neurology* 23 (1969): 11–17.

14. See R. W. Sperry, "Brain bisection and mechanisms of consciousness," in J. C. Eccles, ed., *Brain and Conscious Experience* (Heidelberg: Springer-Verlag, 1966), pp. 299–313.

15. M. S. Gazzaniga, "Understanding layers: From neuroscience to human responsibility," in A. Battro, S. Dehaene, and W. Singer, eds., *Proceedings of the Working Group on Neurosciences and the Human Person: New Perspectives on Human Activities, Scripta Varia* 121 (Vatican City: Ex Aedibus Academicis, 2013).

16. Op-ed, *Los Angeles Times*, May 18, 1967.

CHAPTER 4: UNMASKING MORE MODULES

1. N. M. Weidman, *Constructing Scientific Psychology: Karl Lashley's Mind-Brain Debates* (Cambridge: Cambridge University Press, 1999).

2. M. S. Gazzaniga, *The Bisected Brain* (New York: Appleton-Century-Crofts, 1970).

3. J. Didion, "Letters from 'Manhattan,'" *New York Review of Books*, August 16, 1979, pp. 18–19.

4. M. S. Gazzaniga, "Lunch with Leon (Festinger)," *Perspectives on Psychological Science* 1 (2006): 88–94.

5. R. G. Collingwood, *An Autobiography* (Oxford: Oxford University Press, 1939).

6. K. Lewin, "1963 Frontiers in group dynamics," in D. Cartwright, ed., *Field Theory in Social Science: Selected Theoretical Papers* (London: Tavistock, 1947), pp. 188–237.

7. L. Festinger, H. Riecken, and S. Schachter, *When Prophecy Fails* (Minneapolis: University of Minnesota Press, 1956).

8. M. S. Gazzaniga, I. S. Szer, and A. M. Crane, "Modification of drinking behavior in the adipsic rat," *Experimental Neurology* 42 (1974): 483–89.

9. M. S. Gazzaniga, "Brain lesions and behavior," in C. Blakemore and M. S. Gazzaniga, eds., *Handbook of Psychobiology* (New York: Academic Press, 1973).

10. D. Premack, "Sameness versus difference: From physical similarity to analogy," 2009, http://www.psych.upenn.edu/~premack/Essays/Entries/2009/5/15_Sameness_Versus_Difference_From_Physical_Similarity_to_Analogy.html.

11. A. Velletri-Glass, M. S. Gazzaniga, and D. Premack, "Artificial language training in global aphasics," *Neuropsychologia* 11 (1973): 95–103.

12. M. S. Gazzaniga, A. Velletri-Glass, M. T. Sarno, and J. B. Posner, "Pure word deafness and hemispheric dynamics: A case history," *Cortex* 9 (1973): 136–43.

13. Ibid.

14. M. S. Gazzaniga, "One brain—two minds?," *American Scientist* 60 (1972): 311–17.

15. D. Hume, *A Treatise of Human Nature,* ed. L. A. Selby-Bigge (Oxford: Clarendon Press, 1896). (Reprinted from D. Hume, *A Treatise of Human Nature* [London: John Noon, 1739].)

16. "Normative," *Wikipedia,* http://en.wikipedia.org/wiki/Normative.

17. A. R. Gibson and M. S. Gazzaniga, "Hemisphere differences in eating behavior in split-brain monkeys," *Physiologist* 14 (1971): 150.

18. J. D. Johnson and M. S. Gazzaniga, "Reversal behavior in split-brain monkeys," *Physiology and Behavior* 6 (1971): 707–709.

19. J. D. Johnson and M. S. Gazzaniga, "Cortical-cortical pathways involved in reinforcement," *Nature* 223 (1969): 71.

20. D. G. Deutsch et al., "Analysis of protein levels and synthesis after learning in the split-brain pigeon," *Brain Research* 198 (1980): 135–45.

21. M. S. Gazzaniga, "Interhemispheric communication of visual learning," *Neuropsychologia* 4 (1966): 183–89.

22. D. H. Wilson, A. G. Reeves, and M. S. Gazzaniga, "'Central' commissurotomy for intractable generalized epilepsy," *Neurology* 32 (1982): 687–97.

23. G. Risse, J. E. LeDoux, D. H. Wilson, and M. S. Gazzaniga, "The anterior commissure in man: Functional variation in a multi-sensory system," *Neuropsychologia* 16 (1975): 23–31.

24. J. E. LeDoux, D. H. Wilson, and M. S. Gazzaniga, "Block design performance following callosal sectioning: Observations on functional recovery," *Archives of Neurology* 35 (1978): 506–508.

25. J. LeDoux, *The Cognitive Neuroscience of Mind: A Tribute to Michael S. Gazzaniga* (Cambridge, MA: MIT Press, 2010).

26. M. S. Gazzaniga, J. E. LeDoux, C. S. Smylie, and B. T. Volpe, "Plasticity in speech organization following commissurotomy," *Brain* 102 (1979): 805–15.

CHAPTER 5:
BRAIN IMAGING CONFIRMS SPLIT-BRAIN SURGERIES

1. B. Volpe, J. LeDoux, and M. Gazzaniga, "Information processing on visual stimuli in an extinguished field," *Nature* 282 (1979): 722–24.

2. L. Weiskrantz, *Blindsight: A Case Study and Implications* (Oxford: Oxford University Press, 1986).

3. J. Holtzman, "Interactions between cortical and subcortical visual areas:

Evidence from human commissurotomy patients," *Vision Research* 24, no. 8 (1984): 801–14.

4. S. M. Kosslyn, J. D. Holtzman, M. J. Farah, and M. S. Gazzaniga, "A computational analysis of mental image generation: Evidence from functional dissociations in split-brain patients," *Journal of Experimental Psychology: General* 114 (1985): 311–41.

5. Pierre S. DuPont addressing the French National Assembly in 1790.

6. G. A. Miller, *Language and Communication* (New York: McGraw-Hill, 1951).

7. N. Chomsky, *Syntactic Structures* (New York: Mouton, 1957).

8. G. A. Miller and N. Chomsky (1963). "Finitary models of language users," in G. A. Miller & N. Chomsky, eds., *Handbook of Mathematical Psychology* (New York: Wiley, 1963), pp. 421–91.

9. G. A. Miller, "The cognitive revolution: A historical perspective," *Trends in Cognitive Science* 7, no. 3 (2003): 141–44.

10. J. D. Watson and F. H. C. Crick, "A structure for deoxyribose nucleic acid," *Nature* 171 (1953): 737–38.

11. J. D. Holtzman, J. J. Sidtis, B. T. Volpe, D. H. Wilson, and M. S. Gazzaniga, "Dissociation of spatial information for stimulus localization and the control of attention," *Brain* 104 (1981): 861–72.

12. J. R. Moeller, B. T. Volpe, J. S. Perlmutter, M. E. Raichle, and M. S. Gazzaniga, "Brain pattern space: A new analytic method uncovers covarying regional values in PET measured patterns of human brain activity," *Society for Neuroscience Abstracts* (1985).

13. M. S. Gazzaniga, *The Social Brain* (New York: Basic Books, 1985).

CHAPTER 6: STILL SPLIT

1. R. Galambos and S. A. Hillyard, *Electrophysiological Approaches to Human Cognitive Processing* (Cambridge, MA: MIT Press, 1981).

2. G. R. Mangun and S. A. Hillyard, "Spatial gradients of visual attention: Behavioral and electrophysiological evidence," *Electroencephalography and Clinical Neurophysiology* 70 (1988): 417–28.

3. N. Jerne, "Antibodies and learning: Selection versus instruction," in G. C. Quarton, T. Melnechuk, and F. O. Schmitt, eds., *The Neurosciences: A Study Program* (New York: Rockefeller University Press, 1967), pp. 200–205.

4. S. Pinker, *The Language Instinct: The New Science of Language and Mind* (New York: William Morrow, 1994).

5. M. S. Gazzaniga, *Nature's Mind* (New York: Basic Books, 1992).

6. R. Granger, J. Ambros-Ingerson, and G. Lynch, "Derivation of encoding characteristics of layer II cerebral cortex," *Journal of Cognitive Neuroscience* 1, no. 1 (1989): 61–87.

7. S. A. Seymour, P. A. Reuter-Lorenz, and M. S. Gazzaniga, "The disconnection syndrome: Basic findings reaffirmed," *Brain* 117 (1994): 105–15.

8. D. M. MacKay and V. MacKay, "Explicit dialog between left and right half-systems of split brains," *Nature* 295 (1982): 690–91.

9. J. Sergent, "Unified response to bilateral hemispheric stimulation by a split-brain patient," *Nature* 305 (1983): 800–802.

10. J. Sergent, "Interhemispheric integration of conflicting information by a split-brain man," *Dyslexia: A Global Issue* 18 (1984): 533–46.

11. See, for example, http://en.wikipedia.org/wiki/Abigail_and_Brittany_Hensel.

12. Abigail and Brittany, http://www.tlc.com/tv-shows/abby-and-brittany.

13. M. S. Gazzaniga, J. D. Holtzman, and C. S. Smylie, "Speech without conscious awareness," *Neurology* 37 (1987): 682–85.

14. S. A. Hillyard and M. Kutas, "Electrophysiology of cognitive processing," *Annual Review of Psychology* 34 (1983): 33–61.

15. Personal communication. Also, S. J. Luck, S. A. Hillyard, G. R. Mangun, and M. S. Gazzaniga, "Independent hemispheric attentional systems mediate visual search in split brain patients," *Nature* 342 (1989): 543–45.

16. J. D. Holtzman, J. J. Sidtis, B. T. Volpe, D. H. Wilson, and M. S. Gazzaniga, "Dissociation of spatial information for stimulus localization and the control of attention," *Brain* 104 (1981): 861–72.

17. P. A. Reuter-Lorenz, G. Nozawa, M. S. Gazzaniga, and H. H. Hughes, "The fate of neglected targets: A chronometric analysis of redundant target effects in the bisected brain," *Journal of Experimental Psychology, Human Perception and Performance* 21 (1995): 211–23.

18. J. D. Holtzman and M. S. Gazzaniga, "Dual task interactions due exclusively to limits in processing resources," *Science* 218 (1982): 1325–27.

19. J. D. Holtzman and M. S. Gazzaniga, "Enhanced dual task performance following callosal commissurotomy in humans," *Neuropsychologia* 23 (1985): 315–21.

20. A. Kingstone, J. T. Enns, G. R. Mangun, and M. S. Gazzaniga, "Guided visual search is lateralized in split-brain patients," *Psychological Science* 6 (1995): 118–21.

21. J. S. Oppenheim, J. E. Skerry, M. J. Tramo, and M. S. Gazzaniga, "Magnetic resonance imaging morphology of the corpus callosum in monozygotic twins," *Annals of Neurology* 26 (1989): 100–104.

22. P. M. Thompson et al., "Genetic influences on brain structure," *Nature Neuroscience* 4 (2001): 1253–58.

23. M. S. Gazzaniga and H. Freedman, "Observations on visual processes after posterior callosal section," *Neurology* 23 (1973): 1126–30.

24. B. T. Volpe, J. J. Sidtis, J. D. Holtzman, D. H. Wilson, and M. S. Gazzaniga, "Cortical mechanisms involved in praxis: Observations following partial and complete section of the corpus callosum in man," *Neurology* 32 (1982): 645–50.

25. See video 7.

26. J. J. Sidtis, B. T. Volpe, J. D. Holtzman, D. H. Wilson, and M. S. Gazzaniga, "Cognitive interaction after staged callosal section: Evidence for a transfer of semantic activation," *Science* 212 (1981): 344–46.

27. M. S. Gazzaniga and C. S. Smylie, "Hemispheric mechanisms controlling voluntary and spontaneous facial expressions," *Journal of Cognitive Neuroscience* 2 (1990): 239–45.

CHAPTER 7: THE RIGHT BRAIN HAS SOMETHING TO SAY

1. S. A. Hillyard and G. R. Mangun, "The neural basis of visual selective attention: A commentary on Harter and Aine," *Biological Psychology* 23, no. 3 (1986): 265–79.

2. G. R. Mangun et al., "Monitoring the visual world: Hemispheric asymmetries and subcortical processes in attention," *Journal of Cognitive Neuroscience* 6 (1994): 265–73.

3. J. C. Eliassen, K. Baynes, and M. S. Gazzaniga, "Anterior and posterior callosal contributions to simultaneous bimanual movements of the hands and fingers," *Brain* 123, no. 12 (2000): 2501–11.

4. http://www.bbc.co.uk/programmes/b01s5b2d.

5. M. S. Gazzaniga, J. D. Holtzman, and C. S. Smylie, "Speech without conscious awareness," *Neurology* 37 (1987): 682–85.

6. K. Baynes and M. S. Gazzaniga, "Right hemisphere language: Insights into normal language mechanisms?," in F. Plum, ed., *Language Communication and the Brain* (New York: Raven Press, 1987).

7. M. S. Gazzaniga et al., "Collaboration between the hemispheres of a callosotomy patient: Emerging right hemisphere speech and the left hemisphere interpreter," *Brain* 119 (1996): 1255–62.

8. M. Kutas, S. A. Hillyard, and M. S. Gazzaniga, "Processing of semantic anomaly by right and left hemispheres of commissurotomy patients: Evidence from event-related potentials," *Brain* 111 (1988): 553–76.

9. M. S. Gazzaniga, J. E. LeDoux, C. S. Smylie, and B. T. Volpe, "Plasticity in speech organization following commissurotomy," *Brain* 102 (1979): 805–15.

10. E. Tulving, *Episodic and Semantic Memory* (New York: Academic Press, 1972), pp. 382–402.

11. Michael Miller, personal communication.

12. L. Nyberg, A. R. McIntosh, and E. Tulving, "Functional brain imaging of episodic and semantic memory with positron emission tomography," *Journal of Molecular Medicine* 76 (1998): 48–53.

13. E. Tulving, S. Kapur, F. I. M. Craik, M. Moscovitch, and S. Houle, "Hemispheric encoding/retrieval asymmetry in episodic memory: Positron emission tomography findings," *Proceedings of the National Academy of Science U.S.A.* 91 (1994): 2016–20.

14. A. M. Owen, B. Milner, M. Petrides, and A. C. Evans, "Memory for object-features versus memory for object-location: A positron emission tomography study of encoding and retrieval processes," *Proceedings of the National Academy of Science U.S.A.* 93 (1996): 9212–17; W. M. Kelley et al., "Hemispheric specialization in human dorsal frontal cortex and medial temporal lobe for verbal and non-verbal memory encoding," *Neuron* 20 (1998): 927–36; A. D. Wagner et al., "Material-specific lateralization of prefrontal activation during episodic encoding and retrieval," *Neuroreport* 1219 (1998): 3711–17; M. B. Miller, A. F. Kingstone, and M. S. Gazzaniga, "Hemispheric encoding asymmetry is more apparent than real," *Journal of Cognitive Neuroscience* 14 (2002): 702–708.

15. D. Zaidel and R. W. Sperry, "Memory impairment after commissurotomy in man," *Brain* 97 (1974): 263–72; E. A. Phelps, W. Hirst, and M. S. Gazzaniga, "Deficits in recall following partial and complete commissurotomy," *Cerebral Cortex* 1 (1991): 492–98.

16. J. E. LeDoux, G. Risse, S. Springer, D. H. Wilson, and M. S. Gazzaniga, "Cognition and commissurotomy," *Brain* 110 (1977): 87–104; J. Metcalfe, M. Funnell, and M. S. Gazzaniga, "Right-hemisphere superiority: Studies of a split-brain patient," *Psychological Science* 6 (1995): 157–63.

17. M. S. Gazzaniga and M. B. Miller, "Testing Tulving: The split brain

approach," in E. Tulving et al., eds., *Memory, Consciousness, and the Brain: The Tallinn Conference* (Philadelphia: Psychology Press, 2000), pp. 307–18.

18. M. S. Gazzaniga, ed., *The New Cognitive Neurosciences,* 2nd ed. (Cambridge, MA: MIT Press, 2000).

CHAPTER 8: STATELY LIVING AND A CALL TO SERVICE

1. L. Thomas, "To Err Is Human," in *The Medusa and the Snail: More Notes of a Biology Watcher* (New York: Viking Press, 1974).

2. M. B. Miller, A. Kingstone, P. M. Corballis, J. Groh, and M. S. Gazzaniga. "Manipulating encoding of faces and associated brain activations," *Society for Neuroscience Abstracts* 25, no. 1 (1999): 646.

3. C. R. Gallistel, *The Organization of Learning* (Cambridge, MA: Bradford Books/MIT Press, 1990).

4. C. R. Hamilton and B. A. Brody, "Separation of visual functions with the corpus callosum of monkeys," *Brain Research* 49 (1973): 15–189.

5. M. S. Gazzaniga, M. Kutas, C. Van Petten, and R. Fendrich, "Human callosal function: MRI verified neuropsychological functions," *Neurology* 39 (1989): 942–46.

6. P. M. Corballis, S. J. Inati, M. G. Funnell, S. Grafton, and M. S. Gazzaniga, "MRI assessment of spared fibers following callosotomy: A second look," *Neurology* 57 (2001): 1345–46.

7. J. D. Van Horn and M. S. Gazzaniga, "Why share data? Lessons learned from the fMRIDC," *Neuroimage* 82 (2013): 677–82.

8. Conan O'Brien's 2011 Dartmouth College Commencement Address, http://www.youtube.com/watch?v=KmDYXaaT9sA.

9. M. G. Funnell, P. M. Corballis, and M. S. Gazzaniga, "A deficit in perceptual matching in the left hemisphere of a callosotomy patient," *Neuropsychologia* 37 (1999): 1143–54.

10. A. Baird, J. Fugelsang, and C. Bennett, "'What were you thinking?': An fMRI study of adolescent decision making," poster presented at the annual meeting of the Cognitive Neuroscience Society, New York, 2005.

11. A. A. Baird, M. K. Colvin, J. Van Horn, S. Inati, and M. S. Gazzaniga, "Functional connectivity: Integrating behavioral, DTI and fMRI data sets," *Journal of Cognitive Neuroscience* 17, no. 4 (2005): 1–8.

12. M. K. Colvin, M. G. Funnell, and M. S. Gazzaniga, "Numerical pro-

cessing in the two hemispheres: Studies of a split-brain patient," *Brain and Cognition* 57, no. 1 (2005): 43–52.

13. R. Seltzer, *Mortal Lessons: Notes on the Art of Surgery* (New York: Simon & Schuster, 1974).

14. "Academy of Sciences urges ban on human cloning," CNN.com, 2002, http://edition.cnn.com/2002/HEALTH/01/18/academies.cloning/index.html.

15. Transcript, President's Council on Bioethics, February 12, 2002, Mielaender questioning Weissman, http://bioethics.georgetown.edu/pcbe/transcripts/feb02/feb13session2.html.

16. M. S. Gazzaniga, "Zygotes and people aren't quite the same," *New York Times,* April 25, 2002.

17. W. Safire, "The but-what-if factor," *New York Times,* May 7, 2002.

18. "Human cloning and human dignity: An ethical inquiry," President's Council on Bioethics, July 2002, http://bioethics.georgetown.edu/pcbe/reports/cloningreport/execsummary.html.

19. S. G. Stolberg, "Bush's bioethics advisory panel recommends a moratorium, not a ban, on cloning research," *New York Times,* July 11, 2002.

20. G. Meilaender, "Spare embryos: If they're going to die anyway, does that really entitle us to treat them as handy research material?," *Weekly Standard,* August 26, 2002.

21. S. Yamanaka, "Induction of pluripotent stem cells from mouse embryonic and adult fibroblast cultures by defined factors," *Cell* 126, no. 4 (2006): 663–76.

22. S. Pinker, "The stupidity of dignity. Conservative bioethics' latest, most dangerous ploy," *New Republic,* May 28, 2008.

CHAPTER 9:
LAYERS AND DYNAMICS: SEEKING NEW PERSPECTIVES

1. D. Kahneman, *Thinking, Fast and Slow* (New York: Farrar, Straus & Giroux, 2011).

2. G. A. Miller, "The magical number seven, plus or minus two: Some limits on our capacity for processing information," *Psychological Review* 63, no. 2 (1956): 81–97.

3. C. Sherrington, *Man on His Nature* (Cambridge: Cambridge University Press, 1940).

4. R. Sperry, "The functional results of muscle transposition in the hind limb of the rat," *Journal of Comparative Neurology* 73, no. 3 (1939): 379–404.

5. R. Sperry, "Functional results of crossing sensory nerves in the rat," *Journal of Comparative Neurology* 78, no. 1 (1943): 59–90.

6. J. Topál, G. Gergely, A. Erdöhegyi, G. Csibra, and A. Miklosi, "Differential sensitivity to human communication in dogs, wolves, and human infants," *Science* 325 (2009): 1269–72.

7. G. Csibra and G. Gergely, "Social learning and social cognition: The case for pedagogy," in Y. Munakata and M. H. Johnson, eds., *Processes of Change in Brain and Cognitive Development: Attention and Performance XXI* (Oxford: Oxford University Press, 2006), pp. 249–74.

8. N. Kapur, T. Manly, J. Cole, and A. Pascual-Leone, *The Paradoxical Brain—So What?* (Cambridge: Cambridge University Press, 2011).

9. J. B. Clarke and L. Sokoloff, "Circulation and energy metabolism of the brain," in G. J. Siegel et al., eds., *Basic Neurochemistry*, 6th ed. (Philadelphia: Lippincott-Raven, 1999), pp. 637–69.

10. M. Kirschner and J. Gerhart, "Evolvability," *Proceedings of the National Academy of Science* 95, no. 15 (1998): 8420–27.

11. Andy Clark, Sage Lecture Series, University of California, Santa Barbara, 2011.

12. M. Rayport, S. Sani, and S. M. Ferguson, "Olfactory gustatory responses evoked by electrical stimulation of amygdalar region in man are qualitatively modifiable by interview content: Case report and review," *International Review of Neurobiology* 76 (2006): 35–42.

13. J. Goldstein, "Emergence as a construct: History and issues," *Emergence: Complexity and Organization* 1, no. 1 (1999): 49–72.

14. P. A. Anderson, "More is different," *Science* 177 (1972): 393–96.

15. R. Sperry, "Brain bisection and mechanisms of consciousness," in J. C. Eccles, ed., *Brain and Conscious Experience* (New York: Springer-Verlag, 1966), pp. 298–313.

16. Sara Bernal, personal communication.

17. D. Davidson, "Mental Events," in L. Foster and J. W. Swanson, eds., *From Experience and Theory* (Amherst: University of Massachusetts Press, 1970), pp. 9–101.

18. D. K. Lewis, *On the Plurality of Worlds* (Oxford: Blackwell, 1986).

19. E. P. Hoel, L. Albantakis, and G. Tononi, "When macro beats micro: Quantifying causal emergence," *Proceedings of the National Academy of Sciences* (in press).

20. *Time,* March 28, 2001.

21. G. Ross, "An interview with Marc Kirschner and John Gerhart,"

American Scientist 100, no. 5 (2013), retrieved August 22, 2013, from http://www.americanscientist.org/bookshelf/pub/marc-kirschner-and-john-gerhart.

APPENDIX I

1. M. S. Gazzaniga, "1981 Nobel prize for physiology or medicine," *Science* 214, no. 4520 (1981): 517–20.

FIGURE CREDITS

VIDEO FIGURES

All links current as of publication.

CHAPTER 2: DISCOVERING A MIND DIVIDED

Video 1: https://vimeo.com/96626442
Early documentary on split-brain work where I was asked to describe how we tested the early split-brain patients. Believe it or not, I was old enough to shave. The experimental set-up at the time of this filming was a step up from the original back-projection screen that was hanging from an exposed pipe in one of the lab rooms in Alles Hall.

Video 2: https://vimeo.com/96626444
Case N.G. swimming not long after her surgery and demonstrating the full callosal sectioning seemingly did not disrupt basic bilateral coordination in any way. In short, the untrained observer would be hard-pressed to detect that the two halves of the brain had been surgically separated.

Video 3: https://vimeo.com/96626445

The original film was shot by Baron Wolman a talented young photographer and one of the founders of Rolling Stone magazine. It shows Case W.J. easily putting together four colored blocks with his left hand to match a sample picture provided to him. His left hand gained its major control from the right hemisphere. When the dominant right hand tried, he simply failed. When both were free to try, one seemed to undue the accomplishments of the other.

CHAPTER 3: SEARCHING FOR THE BRAIN'S MORSE CODE

Video 4: https://vimeo.com/96626446

Case D.R. A patient from the East Coast series of cases carrying out command to posture either her left or right hand. Watch a few times and you will begin to see how she uses self-cueing strategies to achieve her goal.

Video 5: https://vimeo.com/96626447

Case N.G. is presented words and pictures exclusively to her right hemisphere. She cannot name them, even though her left hand is able to find the correct object.

Video 6: https://vimeo.com/96627695}

Emotional states spread throughout the brain rapidly. Here Case N.G. was shown evocative nude photos to the right hemisphere. While her left hemisphere couldn't say what the picture had been, it could recognize something funny had happened.

CHAPTER 4: UNMASKING MORE MODULES

Video 7: https://vimeo.com/96627698

Filmed in our original trailer set-up, we would lateralize questions to the right hemisphere by first saying "Who is your favorite____?" and then lateralize the last part of the question to either the right or left hemisphere. Here we asked the right hemisphere "Who is your favorite girlfriend?" Since he was a patient that could control both hands from one hemisphere, both cooperated in arranging Scrabble letters to spell "LIZ."

Video 8: https://vimeo.com/96627699

We examined Case P.S.'s right hemisphere on many dimensions. Here we ask "Who are you?" The right hemisphere answers "Paul."

Video 9: https://vimeo.com/96627700
Case P.S.'s right hemisphere tells us about his favorite TV show and Henry Winkler.

Video 10: https://vimeo.com/96627702
Case J.W. working on a simple experiment a few years later and filmed by Robert Bazell at NBC News. It was "live" science and it all worked.

CHAPTER 6: STILL SPLIT

Video 11: https://vimeo.com/96628407
A double grid with each hemisphere seeing one grid of nine cells. Lights would appear randomly, four at a time, in each field—an overwhelming experience for normal subjects. Split-brain patients could manage the task with ease.

Video 12: https://vimeo.com/96628410
J.W. being examined by me in our GMC Eleganza van. The word "sun" had been flashed to left brain and a black-and-white line drawing of a traffic light to his right brain. Teaching him how to play Twenty Questions also taught him how to gain access to the right-brain information from the left brain.

Video 13: https://vimeo.com/96628408
Case J.W. being instructed to smile from the left hemisphere. Watch the asymmetrical retraction of his face muscles on right side of his face, followed by the left-side response. Also notice the asymmetry as the face starts to regain the neutral posture.

CHAPTER 7: THE RIGHT BRAIN HAS SOMETHING TO SAY

Video 14: https://vimeo.com/96628409
Jim Eliassen's task where J.W. is able to do two things at once whereas most of us cannot.

Note: *Italicized* page numbers refer to picture captions.